T0224587

SpringerBriefs in Energy

SpringerBriefs in Energy presents concise summaries of cutting-edge research and practical applications in all aspects of Energy. Featuring compact volumes of 50 to 125 pages, the series covers a range of content from professional to academic. Typical topics might include:

- A snapshot of a hot or emerging topic
- A contextual literature review
- A timely report of state-of-the art analytical techniques
- An in-depth case study
- A presentation of core concepts that students must understand in order to make independent contributions.

Briefs allow authors to present their ideas and readers to absorb them with minimal time investment.

Briefs will be published as part of Springer's eBook collection, with millions of users worldwide. In addition, Briefs will be available for individual print and electronic purchase. Briefs are characterized by fast, global electronic dissemination, standard publishing contracts, easy-to-use manuscript preparation and formatting guidelines, and expedited production schedules. We aim for publication 8–12 weeks after acceptance.

Both solicited and unsolicited manuscripts are considered for publication in this series. Briefs can also arise from the scale up of a planned chapter. Instead of simply contributing to an edited volume, the author gets an authored book with the space necessary to provide more data, fundamentals and background on the subject, methodology, future outlook, etc.

SpringerBriefs in Energy contains a distinct subseries focusing on Energy Analysis and edited by Charles Hall, State University of New York. Books for this subseries will emphasize quantitative accounting of energy use and availability, including the potential and limitations of new technologies in terms of energy returned on energy invested.

Tim Aschenbruck · Jörg Dickert ·
Willem Esterhuizen · Bartosz Filipecki ·
Sara Grundel · Christoph Helmberg ·
Tobias K. S. Ritschel · Philipp Sauerteig ·
Stefan Streif · Andreas Wasserrab ·
Karl Worthmann

Hierarchical Power Systems: Optimal Operation Using Grid Flexibilities

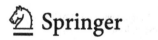 Springer

Tim Aschenbruck
Technische Universität Chemnitz
Chemnitz, Germany

Willem Esterhuizen
Technische Universität Ilmenau
Ilmenau, Germany

Sara Grundel
Max Planck Institute for Dynamics
of Complex Technical Systems
Magdeburg, Saxony-Anhalt, Germany

Tobias K. S. Ritschel
Technical University of Denmark
Kongens Lyngby, Denmark

Stefan Streif
Technische Universität Chemnitz
Chemnitz, Germany

Karl Worthmann
Technische Universität Ilmenau
Ilmenau, Germany

Jörg Dickert
SachsenNetze HS.HD GmbH
Dresden, Germany

Bartosz Filipecki
Università di Pisa
Pisa, Italy

Christoph Helmberg
Technische Universität Chemnitz
Chemnitz, Germany

Philipp Sauerteig
Technische Universität Chemnitz
Chemnitz, Germany

Andreas Wasserrab
TenneT TSO GmbH
Bayreuth, Germany

This work was funded by the Federal Ministry of Education and Research, Germany (BMBF; grants 5M18OCA, 05M18SIA, and 05M18EVA).

ISSN 2191-5520   ISSN 2191-5539   (electronic)
SpringerBriefs in Energy
ISBN 978-3-031-25698-1   ISBN 978-3-031-25699-8   (eBook)
https://doi.org/10.1007/978-3-031-25699-8

This Springer imprint is published by the registered company Springer Nature Switzerland AG
The registered company address is: Gewerbestrasse 11, 6330 Cham, Switzerland

Preface

This book is the result of the project Consistent Optimization aNd Stabilization of Electrical Networked Systems (CONSENS)[1] combining the expertise of several academic research groups from Technische Universität Chemnitz; Max Planck Institute for Complex Dynamical Systems, Magdeburg; Technische Universität Ilmenau as well as the distribution system operator SachsenNetze GmbH and the transmission system operator TenneT TSO GmbH.

We consider the power grid as a hierarchy made up of the transmission, distribution, and microgrid levels. We develop interfaces among these levels and show how flexibility in power demand associated with residential batteries can be communicated through the entire grid to facilitate the optimal power-flow computations within the transmission grid. To realize this approach, we combine model-order reduction on the distribution level with semi-definite optimal power flow on the transmission level. Moreover, we introduce a new heuristic algorithm that attempts to arrive at a stable power-flow set-point in the transmission grid, i.e., where the generators are synchronized. The potential of this grid-wide optimization is shown in a

---

[1] https://konsens.github.io.

numerical case study based on modified IEEE 9-bus and 33-bus systems for transmission and distribution grid, respectively. Our results show that exploiting the flexibility improves the performance of the power flow within the transmission grid.

Chemnitz, Germany   Tim Aschenbruck
Dresden, Germany   Jörg Dickert
Ilmenau, Germany   Willem Esterhuizen
Pisa, Italy   Bartosz Filipecki
Magdeburg, Germany   Sara Grundel
Chemnitz, Germany   Christoph Helmberg
Kongens Lyngby, Denmark   Tobias K. S. Ritschel
Chemnitz, Germany   Philipp Sauerteig
Chemnitz, Germany   Stefan Streif
Bayreuth, Germany   Andreas Wasserrab
Ilmenau, Germany   Karl Worthmann

Contents

Chapter 1
Introduction

New distributed energy storage and generation devices in the distribution grid provide flexibility that the entire power grid could exploit to enhance operation. However, this is hampered by the fact that the grid is divided into hierarchical parts of different voltage levels that are usually operated by independent entities.

Transmission system operators (TSOs) operate extra high voltage grids, which transmit electrical energy from power plants over long distances. Distribution system operators (DSOs) operate high voltage grids, which distribute the transmitted energy over shorter distances to industrial, commercial, and residential customers. The transmission and distribution grids are typically treated as separate entities in power-flow analyses: distribution grids are often treated as fixed loads from a transmission grid's perspective, and transmission grids are often treated as fixed power supplies from a distribution grid's perspective. Therefore, new tools that support the coordination between the TSOs, DSOs and flexibility providers (e.g., smart microgrids) in real-time are becoming more relevant [9].

Many contributions have been made to the coordinated analysis of transmission and distribution grids during the last decade, e.g., for state estimation [26], economic dispatch [18], unit commitment [14], power-flow [25], and optimal power-flow (OPF) [19, 24]. Risk assessment [12] and Nash equilibria [21] of such coupled grids have also been considered. Most research in this area is concerned with distributed algorithms for solving the above problems with minimal computational complexity and data transfer between TSOs and DSOs. We refer to the literature review by Li [17] for more details. Distribution grids with flexible storage and generation can be organized as collections of microgrids whose operation is coordinated using distributed optimal control [15]. Optimal operation of such microgrids [30] is a topic of intensive research. For instance, Dall'Anese et al. [8] and Levron et al. [16] consider optimal power-flow of microgrids (see [1] for a review), and Wu and Guan [27] and Zhang et al. [29] propose optimization algorithms for optimal operation of interconnected microgrids.

© The Author(s), under exclusive license to Springer Nature Switzerland AG 2023
T. Aschenbruck et al., *Hierarchical Power Systems: Optimal Operation Using Grid Flexibilities*, SpringerBriefs in Energy,
https://doi.org/10.1007/978-3-031-25699-8_1

The large-scale nature of power grid models makes them challenging to analyze in real time. Therefore, the subject of identifying small-scale *surrogate* models [2] (also called reduced order models (ROMs) or *equivalents*) which approximate the original models has been an important research topic for several decades and in many different fields. For power grids, and power systems in general, both static [6] and dynamical models [13, 23] have been considered.

A critical aspect of grid operation is to maintain system security [20] and stability [10]. As distribution grids become more active and more energy is produced using renewable energy sources (RESs), the risk of instability and blackouts increases. Therefore, incorporating security and stability constraints into the optimal power-flow problems [5, 28] is becoming ever more relevant. The security constraints are usually steady-state (also called small signal) stability constraints [7, 22], or transient stability constraints [3, 11].

In this book we combine optimization approaches, model order reduction techniques for power-flow problems, security-constrained OPF and a new stability algorithm to make flexibility associated with residential batteries available for the optimal power-flow computations on the transmission grid level. To this end, we develop interfaces among the different levels of the grid hierarchy in order to communicate information from microgrids (MGs) through the DSO to the TSO and back to the MGs. Throughout the book we focus on the technical aspects of exploiting flexibility throughout the grid, and ignore details related to market boundaries/rules between the grid levels.

1.1 Main Idea

Distributed energy storage can be used to improve the operation of microgrids, e.g., through *peak-shaving* of the power demand. While this reduces the volatility of the microgrid power demand, the DSOs and TSOs are not able to further exploit this potential for storing and releasing energy, e.g., when renewable energy production is very high or low. The operation of each individual microgrid can *in principle* be coordinated with that of the DSOs and TSOs by solving one large-scale optimization problem encompassing all involved grids. However, such a problem would be computationally intractable for realistically sized grids and it would require the involved parties to share sensitive data.

Therefore, in this work, we distinguish between three levels, shown in Fig. 1.1, in the hierarchy of the electrical power grid: the MG level (in green), the DSO level (in yellow/orange), and the TSO level (in red), with communication between them taking place via *interface nodes*. In general the DSO level contains the MG level, but in our approach we consider these levels separately.

We communicate the flexibility of each microgrid's power demand to the DSO whose power demand from the TSO is therefore also flexible. Then, the DSO computes the minimal and maximal amount of power the TSO can supply. This allows the TSO to incorporate the microgrid flexibility in its planning. However, in order for the TSO to also take the objectives of the microgrids into account (in addition to its

Fig. 1.1 Hierarchical structure of the German power grid. Flexibility is transported from bottom to top: low-voltage microgrid (MG) level (green), medium and high voltage distribution system operator (DSO) level (yellow/orange), and extra-high voltage transmission system operator (TSO) level (red)

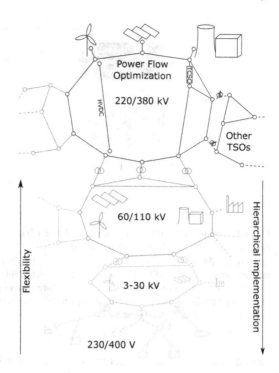

own), it is necessary to compute a *desired* power demand for each microgrid. In this work, we use peak-shaving optimization for this purpose. Then, the DSO computes the amount of power the TSO can supply without violating the limits of the DSO grids. Given this information, the TSO computes an optimal trade-off between, on one hand, the economic cost of operating the power plants and, on the other hand, the economic cost of deviating from the desired power demand of the DSO. Once the TSO knows the amount of power allocated to the DSO, the DSO computes an optimal trade-off between the amount of power delivered to each microgrid, i.e., it attempts to deliver the desired power. This entire process is then repeated at regular time intervals.

Next, we elaborate on the responsibilities of each level in our approach and in particular, we comment on the use of semidefinite programming, security- and synchronization constraints, and model reduction.

1.1.1 Microgrid Level

At the MG level (Chap. 3) residential energy storage devices such as batteries are used to manipulate the active power demand. For instance, charging the batteries, which can be seen as an artificial load, increases the active power demand while discharging

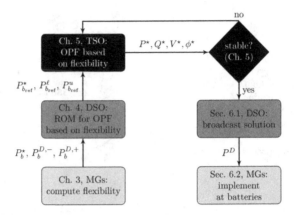

Fig. 1.2 Outline of the book. First, lower and upper bounds as well as an optimum (in terms of peak shaving) of the active power demand per microgrid (MG) are computed to determine the range of flexibility $[P_b^{D,-}, P_b^{D,+}]$ and associated costs. Then, the DSO uses ROMs to broadcast this information to the TSO, where a power-flow set-point, which also results in synchronized generators, is found via a heuristic iterative procedure. Then a power-flow problem is solved on the DSO level with the supplied power fixed at the interface between TSO and DSO. Finally, the power delivered is distributed among the MGs

them decreases it. Thus, the active power demand within each MG can be modified subject to battery constraints. In particular, the batteries yield a feasible range for the active power demand with lower and upper bound. Within this range the active power demand is flexible. Furthermore, an optimal control strategy of the batteries for *peak shaving* is computed, i.e. to reduce the volatility of the net power consumption. The resulting active power demand is used to associate costs with the amount of flexibility. Both the flexibility limits as well as the optimal power demand are communicated to the DSO level. Then, the DSO and TSO levels determine what the active power demand of the MGs should be to, for example, help address congestions, and each MG distributes this power among the residential batteries (Chap. 6).

1.1.2 DSO Level

The DSO level in our approach represents the intermediate level between the MGs and the TSO level (Chap. 4). Here, the main task is to provide active power limits as well as an optimal active power value to the TSO level based on the flexibilities provided by the MGs. For this purpose, we solve two optimization problems (essentially, modified OPF problems) where the complex MG power demands are considered flexible. The objectives are to minimize and maximize the active power provided by the transmission grid. Furthermore, we solve the power-flow equations to determine the amount of active power drawn from the transmission grid for the optimal active MG power demands (we refer to this as the *optimal* active power required from the transmission grid).

We solve the optimization problems using semidefinite programming (SDP), i.e., by using semidefinite relaxations [33]. The advantage of this approach is that (1) unlike the original problem, the resulting SDP is convex and has a unique solution, and (2) it is more accurate than using linearization. It is particularly well-suited for radial grids (i.e., grids that have tree structure, distribution grids are typically radial) because the relaxation yields the exact solution to the original problem under mild conditions [4]. Furthermore, we use a clustering approach to create a surrogate model of the distribution grid which contains significantly fewer buses than the original grid. Using this surrogate model, we can solve the optimization problems more efficiently. We compare the solutions obtained using SDP and the surrogate model to those obtained using a standard nonlinear programming approach. We use MATPOWER 7.1 [31, 32] for this purpose.

The TSO computes a stable optimal power-flow solution, taking the minimal, maximal and optimal active powers demanded by the DSO level into account. This determines the amount of power allocated to the distribution grid. Finally, after this power-flow solution has been determined by the TSO, we solve another modified OPF problem on the DSO level to compute the corresponding active power provided to each of the MGs. Here, the objective is to minimize the sum of the squared deviations of the active MG power demands from their optimal values.

1.1.3 TSO Level

The goal on the TSO level is to solve the optimal power-flow problem with the communicated power limits as well as the optimal power for the slack nodes representing the interface to the DSO level (Chap. 5). The optimal solution must remain within the given bounds for flexible nodes and additionally the deviation from the communicated optimal value at a flexible node is penalized via quadratic costs. We solve two separate problems on the TSO level: the security-constrained OPF, which ensures that the solution is secure to possible line failures, and the stability-guaranteed power-flow problem, which ensures that a power-flow set-point results in synchronized generators. In this latter problem the OPF problem is iteratively invoked in a heuristic algorithm to arrive at a power-flow set-point which also corresponds to synchronized machines. We explain the interface between the static power-flow problem and the dynamic synchronization problem in Sect. 5.2. The result is then communicated downwards to the MG level via the DSO level as explained in previous sections.

1.2　Outline

The remainder of the book is structured as depicted in Fig. 1.2. In Chap. 2, we describe the power-flow equations, the solution of optimal power-flow problems using semidefinite programming, and the structure-preserving model used in the stability analysis. Next, we analyze what amount of flexibility the MGs can provide and what data is required on the DSO level in Chap. 3. Then, in Chap. 4, the DSO broadcasts this information to the TSO level using ROMs in order to solve the OPF problems efficiently. The TSO determines the optimal power demands in Chap. 5 and checks the solution to the security-constrained OPF in Sect. 5.1, and for stability in Sect. 5.2. Finally, the optimal power demands are broadcast through the DSO level to the MGs and implemented at the residential batteries in Chap. 6. Chapter 7 concludes the book with a detailed numerical example implemented on modified IEEE 9-bus and 33-bus systems.

References

1. Abdi H, Beigvand SD, La Scala M (2017) A review of optimal power-flow studies applied to smart grids and microgrids. Renew Sustain Energy Rev 71:742–766
2. Antoulas AC (2005) Approximation of large-scale dynamical systems, vol 6. In: Advances in design and control. SIAM Publications, Philadelphia, PA
3. Aschenbruck T, Esterhuizen W, Streif S (2020) Transient stability analysis of power grids with admissible and maximal robust positively invariant sets. Automatisierungstechnik 68(12):1011–1021
4. Bose S, Gayme DF, Low S, Chandy KM (2011) Optimal power-flow over tree networks. In: 2011 49th annual Allerton conference on communication, control, and computing (Allerton), pp 1342–1348
5. Capitanescu F, Ramos JLM, Panciatici P, Kirschen D, Marcolini AM, Platbrood L, Wehenkel L (2011) State-of-the-art, challenges, and future trends in security constrained optimal power flow. Electr Power Syst Res 81:1731–1741
6. Chow JH (ed) (2013) Power system coherency and model reduction. In: Power electronics and power systems, vol 94. Springer
7. Cui B, Sun XA (2018) A new voltage stability-constrained optimal power-flow model: sufficient condition, SOCP representation, and relaxation. IEEE Trans Power Syst 33(5):5092–5102
8. Dall'Anese E, Zhu H, Giannakis GB (2013) Distributed optimal power flow for smart microgrids. IEEE Trans Smart Grid 4(3):1464–1475
9. ENTSO-E. Vision on market design and system operation towards 2030. https://vision2030.entsoe.eu. Accessed 15 Feb 2021
10. Gajduk A, Todorovski M, Kocarev L (2014) Stability of power grids: an overview. Eur Phys J Spec Top 223:2387–2409
11. Geng G, Abhyankar S, Wang X, Dinavahi V (2017) Solution techniques for transient stability-constrained optimal power flow—part II. IET Gener Transm Distrib 11(12):3186–3193
12. Jia H, Qi W, Liu Z, Wang B, Zeng Y, Xu T (2015) Hierarchical risk assessment of transmission system considering the influence of active distribution network. IEEE Trans Power Syst 30(2):1084–1093
13. Joo S-K, Liu C-C, Jones LE, Choe J-W (2004) Coherency and aggregation techniques incorporating rotor and voltage dynamics. IEEE Trans Power Syst 19(2):1068–1075

14. Kargarian A, Fu Y (2014) System of systems based security-constrained unit commitment incorporating active distribution grids. IEEE Trans Power Syst 29(5):2489–2498
15. Lasseter RH (2011) Smart distribution: coupled microgrids. Proc IEEE 99(6):1074–1082
16. Levron Y, Guerrero JM, Beck Y (2013) Optimal power flow in microgrids with energy storage. IEEE Trans Power Syst 28(3):3226–3234
17. Li Z (2018) Distributed transmission-distribution coordinated energy management based on generalized master-slave splitting theory. In: Springer theses. Springer
18. Li Z, Guo Q, Sun H, Wang J (2016) Coordinated economic dispatch of coupled transmission and distribution systems using heterogeneous decomposition. IEEE Trans Power Syst 31(6):4817–4830
19. Li Z, Guo Q, Sun H, Wang J (2018) Coordinated transmission and distribution AC optimal power flow. IEEE Trans Smart Grid 9(2):1228–1240
20. Mithun M, Srinivas M, Sydulu M (2010) Security constraint optimal power flow (SCOPF)—a comprehensive survey. Int J Comput Appl 11
21. Mezghani I, Papavasiliou A, Le Cadre H (2018) A generalized Nash equilibrium analysis of electric power transmission-distribution coordination. In: Proceedings of the 9th ACM international conference on future energy systems, Karlsruhe, Germany, June 2018, pp 526–531
22. Milano F, Cañizares CA, Invernizzi M (2005) Voltage stability constrained OPF market models considering $N-1$ contingency criteria. Electr Power Syst Res 74:27–36
23. Milano F, Srivastava K (2009) Dynamic REI equivalents for short circuit and transient stability analyses. Electr Power Syst Res 79(6):878–887
24. Mohammadi A, Mehrtash M, Kargarian A (2019) Diagonal quadratic approximation for decentralized collaborative TSO+DSO optimal power flow. IEEE Trans Smart Grid 10(3):2358–2370
25. Sun H, Guo Q, Zhang B, Guo Y, Li Z, Wang J (2015) Master-slave-splitting based distributed global power flow method for integrated transmission and distribution analysis. IEEE Trans Smart Grid 6(3):1484–1492
26. Sun HB, Zhang BM (2005) Global state estimation for whole transmission and distribution networks. Electr Power Syst Res 74:187–195
27. Wu J, Guan X (2013) Coordinated multi-microgrids optimal control algorithm for smart distribution management system. IEEE Trans Smart Grid 4(4):2174–2181
28. Xu Y, Dong ZY, Xu Z, Zhang R, Wong KP (2012) Power system transient stability-constrained optimal power flow: a comprehensive review. In: Proceedings of the 2012 IEEE power and energy society general meeting, p 13170106
29. Zhang Y, Xie L, Ding Q (2016) Interactive control of coupled microgrids for guaranteed system-wide small signal stability. IEEE Trans Smart Grid 7(2):1088–1096
30. Zhu Z, Tang J, Lambotharan S, Chin WH, Fan Z (2012) An integer linear programming based optimization for home demand-side management in smart grid. In: Proceedings of the 2012 IEEE PES innovative smart grid technologies (ISGT), p 12650537
31. Zimmerman RD, Murillo-Sánchez CE (2020) MATPOWER (version 7.1)
32. Zimmerman RD, Murillo-Sánchez CE, Thomas RJ (2011) MATPOWER: steady-state operations, planning, and analysis tools for power systems research and education. IEEE Trans Power Syst 26(1):12–19
33. Zorin IA, Gryazina EN (2019) An overview of semidefinite relaxations for optimal power flow problem. Autom Remote Control 80(5):813–833

Chapter 2
Preliminary Theory

In this chapter, we present the power-flow equations [5, 9], the optimal power-flow problem, the semidefinite approach for solving optimal power-flow problems [7], and the dynamic structure-preserving power grid model [8] which are relevant to several of the sections in the remainder of the book.

2.1 Optimal Power-Flow

The optimal power-flow [3] is an optimization problem that aims to satisfy power network constraints while achieving a minimum cost. The constraints include power balance at network nodes, as well as limits on line capacity, voltage and power generation. The objective function can represent, for example, power generation cost or the total amount of power lost during network operation.

The power-flow equations govern the flow of electricity in a power grid according to the Kirchhoff's laws. The grid consists of N buses which are interconnected by a set of transmission lines, each modeled as a *nominal π circuit* [5]. Each bus is characterized by its complex voltage, V_b, complex power generation, S_b, and complex power demand, S_b^D. We represent the voltage in polar form and the powers in rectangular form, i.e.,

$$V_b = |V_b| \exp(\mathbf{i}\phi_b), \quad b \in \mathcal{B}, \tag{2.1a}$$

$$S_b = P_b + \mathbf{i}Q_b, \quad b \in \mathcal{B}, \tag{2.1b}$$

$$S_b^D = P_b^D + \mathbf{i}Q_b^D, \quad b \in \mathcal{B}, \tag{2.1c}$$

© The Author(s), under exclusive license to Springer Nature Switzerland AG 2023
T. Aschenbruck et al., *Hierarchical Power Systems: Optimal Operation Using Grid Flexibilities*, SpringerBriefs in Energy,
https://doi.org/10.1007/978-3-031-25699-8_2

where $|V_b|$ is the voltage magnitude; ϕ_b is the voltage phase angle; $P_b \in \mathbb{R}_{\geq 0}$ and $P_b^D \in \mathbb{R}_{\geq 0}$ are the real power generated and demand, respectively; $Q_b \in \mathbb{R}$ and $Q_b^D \in \mathbb{R}$ are the reactive power generated and demand, respectively; and $\mathbf{i}$ denotes the imaginary unit. Furthermore, $\mathcal{B} = \{1, \ldots, N\}$ indicates the bus indices; $\mathcal{B}_G = \{1, \ldots, N_G\} \subset \mathcal{B}$ indicates the generator indices; and $\mathcal{B}_L = \{N_G, \ldots, N_G + N_L\} \subset \mathcal{B}$ indicates load indices ($N_G + N_L = N$). The decision variables in standard OPF problems are $|V_b|$ and ϕ_b for $b \in \mathcal{B}$; and P_b and Q_b for $b \in \mathcal{B}_G$. That is, in the standard formulation the decision variables are the voltage magnitude and angle at all buses, but the active and reactive power generated at generators only. In this book, we will consider OPF problems on the DSO and TSO level where some of the loads are *flexible*, meaning that the active and reactive powers supplied by them, P_b and Q_b, are additional decision variables in the OPF problem. Details concerning which nodes are flexible and how the standard OPF formulation is modified will be given in the relevant sections.

At each bus, the difference between the generated and consumed power must equal the power supplied from neighboring buses through transmission lines plus what is consumed by shunt-connected elements such as capacitors or inductors and admittance-to-ground:

$$S_b - S_b^D = \sum_{h \in \mathcal{N}_S^{(b)}} S_{bh}^S + \sum_{h \in \mathcal{N}_E^{(b)}} S_{hb}^E + (\tilde{Y}_b)^H |V_b|^2, \quad b \in \mathcal{B}. \tag{2.2}$$

Here, S_{bh}^S denotes the power-flow from bus b into a line starting in b and ending in bus h. Similarly, S_{hb}^E is the power-flow from bus h to a line ending in bus b. The sets $\mathcal{N}_S^{(b)}$ and $\mathcal{N}_E^{(b)}$ contain the indices of buses which are connected to bus b through lines starting and ending in b, respectively. Finally, $\tilde{Y}_b$ represents the bus admittance, and superscript H denotes Hermitian transpose.

The complex line power injections are given by

$$S_{bh}^S = \left(I_{bh}^S\right)^H V_b, \quad h \in \mathcal{N}_S^{(b)}, \quad b \in \mathcal{B}, \tag{2.3a}$$

$$S_{bh}^E = \left(I_{bh}^E\right)^H V_h, \quad h \in \mathcal{N}_S^{(b)}, \quad b \in \mathcal{B}, \tag{2.3b}$$

where I_{bh}^S and I_{bh}^E are the complex current injections at the start and the end of the line going from bus b to bus h. These currents are given by

$$I_{bh}^S = \left(Y_{bh} + Y_{bh}^{SH}\right) V_b - Y_{bh} V_h, \quad h \in \mathcal{N}_S^{(b)}, \quad b \in \mathcal{B}, \tag{2.4a}$$

$$I_{bh}^E = \left(Y_{bh} + Y_{bh}^{SH}\right) V_h - Y_{bh} V_b, \quad h \in \mathcal{N}_S^{(b)}, \quad b \in \mathcal{B}, \tag{2.4b}$$

where Y_{bh} is the admittance of the line and Y_{bh}^{SH} is the shunt capacitance.

In addition to the above equality constraints, there are several operational requirements of the power grids that are expressed as inequalities. First, at each end of every line, there exists a thermal limit T^+ on its capacity to carry current:

$$\left| I_{bh}^{S} \right| \leq T_{bh}^{+}, \quad h \in \mathcal{N}_{S}^{(b)}, \quad b \in \mathcal{B}, \tag{2.5a}$$

$$\left| I_{bh}^{E} \right| \leq T_{bh}^{+}, \quad h \in \mathcal{N}_{S}^{(b)}, \quad b \in \mathcal{B}. \tag{2.5b}$$

Moreover, the voltage magnitude at each node b is bounded between V_{b}^{-} and V_{b}^{+}:

$$V_{b}^{-} \leq |V_{b}| \leq V_{b}^{+}, \quad b \in \mathcal{B}, \tag{2.6}$$

and power generated at a node is constrained as follows:

$$P_{b}^{-} \leq P_{b} \leq P_{b}^{+}, \quad b \in \mathcal{B}, \tag{2.7a}$$

$$Q_{b}^{-} \leq Q_{b} \leq Q_{b}^{+}, \quad b \in \mathcal{B}. \tag{2.7b}$$

We will consider different cost functions (though always convex) in the OPF problem on the DSO and TSO levels. Details on the cost will be given in the relevant sections.

In general, the above gives a full description of the standard optimal power-flow problem. The constraints are, in general, nonlinear and nonconvex, which makes it difficult to solve the problem to global optimality in practice. To address this, in the following section, we consider the SDP relaxation.

2.2 Semidefinite Optimal Power-Flow

Semidefinite programming has been well established as a possible solution method for the optimal power-flow that would be a middle ground between solution efficiency and quality, offered by linear approximations and nonlinear problems respectively [7]. This is especially the case for tree networks, such as distribution systems, where under mild conditions, we can guarantee global optimality of the solution [2, 4]. However, if the conditions are not met, this method can lead to solutions that are infeasible to the original problem [6]. Therefore, caution is required when interpreting the results of the relaxed OPF problem.

To this end, we consider a standard SDP relaxation of the OPF, in which we keep the objective function. We take a vector of voltages $V = (V_1, V_2, \ldots, V_N)^T \in \mathbb{C}^N$ and define a new variable matrix $W = VV^H$, the entries of which replace their corresponding entries in the OPF constraints. The matrix W is Hermitian, positive semidefinite and rank 1 and we drop the latter nonconvex rank constraint to obtain a convex (SDP) optimization problem with decision variables including the matrix W and complex power generation S.

The constraints of the relaxation can be described as follows:

$$S_{bh}^{S} = \left(Y_{bh} + Y_{bh}^{SH} \right)^{H} W_{bb} - Y_{bh}^{H} W_{bh}, \quad h \in \mathcal{N}_{S}^{(b)}, \quad b \in \mathcal{B}, \tag{2.8a}$$

$$S_{bh}^{E} = \left(Y_{bh} + Y_{bh}^{SH} \right)^{H} W_{hh} - Y_{bh}^{H} W_{hb}, \quad h \in \mathcal{N}_{S}^{(b)}, \quad b \in \mathcal{B}, \tag{2.8b}$$

are the complex power injections into the lines as explained in (2.3) and (2.4) with voltage terms replaced by the corresponding entries of matrix W. Then,

$$S_b - S_b^D = \sum_{h \in \mathcal{N}_S^{(b)}} S_{bh}^S + \sum_{h \in \mathcal{N}_E^{(b)}} S_{hb}^E + (\tilde{Y}_b)^H W_{bb}, \quad b \in \mathcal{B}, \tag{2.9}$$

corresponds directly to (2.2), the power flow balance equations, with fixed bus demand S_b^D and variable generation S_b. The thermal line limits (2.5) are expressed in terms of W:

$$\left|I_{bh}^S\right|^2 = \left|Y_{bh} + Y_{bh}^{SH}\right|^2 W_{bb} + |Y_{bh}|^2 W_{hh} - 2\mathrm{Re}\left((Y_{bh} + Y_{bh}^{SH})Y_{bh}^H W_{bh}\right) \tag{2.10a}$$
$$\leq (T_{bh}^+)^2, \quad h \in \mathcal{N}_S^{(b)}, \quad b \in \mathcal{B},$$

$$\left|I_{bh}^E\right|^2 = \left|Y_{bh} + Y_{bh}^{SH}\right|^2 W_{hh} + |Y_{bh}|^2 W_{bb} - 2\mathrm{Re}\left((Y_{bh} + Y_{bh}^{SH})Y_{bh}^H W_{hb}\right) \tag{2.10b}$$
$$\leq (T_{bh}^+)^2, \quad h \in \mathcal{N}_S^{(b)}, \quad b \in \mathcal{B}.$$

Finally, we rephrase the voltage magnitude bounds (2.6) in a similar fashion:

$$(V_b^-)^2 \leq W_{bb} \leq (V_b^+)^2, \quad b \in \mathcal{B}, \tag{2.11}$$

and also include the power generation limits

$$P_b^- \leq P_b \leq P_b^+, \quad b \in \mathcal{B}, \tag{2.12a}$$

$$Q_b^- \leq Q_b \leq Q_b^+, \quad b \in \mathcal{B}, \tag{2.12b}$$

and semidefiniteness constraint on the matrix W:

$$W \succeq 0. \tag{2.13}$$

2.3   Dynamic Structure Preserving Power Grid Model

In Sect. 5.2 we will consider the stability-guaranteed power-flow problem, concerned with finding a power-flow set-point of the TSO grid that results in synchronized connected oscillators modeled by the well-known *structure-preserving* (SP) model, originally introduced in [1]. We now present a summary of this model's derivation, as detailed in [8], emphasizing how the power-flow set-point on the TSO grid determines the constants appearing in the dynamic model.

As introduced in Sect. 2.1, consider the TSO power grid modeled as a graph of $N \in \mathbb{N}$ nodes, and let $N_G \in \mathbb{N}$ and $N_L \in \mathbb{N}$ denote the number of generator and load buses, respectively. In the study of power system stability one is interested in the motion of connected oscillators initiating from an initial state where the power-flow is balanced. In other words, it is assumed that the oscillators are initially operating at an equilibrium point associated with a power-flow set-point, which we label $(P_b^\star, Q_b^\star, P_b^{D\star}, Q_b^{D\star}, |V_b^\star|, \phi_b^\star), b \in \{1, 2, \ldots, N\}$.

The graph is now augmented with N_G *additional load nodes* (called *terminal* nodes) that are added between the generator nodes and the nodes they connect to, which brings the total number of nodes to $\hat{N} := N + N_G$. *These terminal load nodes do not demand or produce power.* The indexing is then changed so that $\mathcal{B}_G = \{1, \ldots, N_G\}$ (as before), $\mathcal{B}_L^{\text{term}} = \{N_G + 1, \ldots, 2N_G\}$ (the terminal load nodes) and $\mathcal{B}_L^{\text{new}} = \{2N_G + 1, \ldots, \hat{N}\}$ (the original load nodes with *new* indexing). The nodes $\mathcal{B}_G$ are referred to as the *internal* generator nodes. We let $Y^{\text{SP}} \in \mathbb{R}^{\hat{N} \times \hat{N}}$ ("SP" for "structure-preserving") denote the square admittance matrix of this new graph, see [8] for details of how this matrix is formed.

Electrically, generators are each represented by a voltage source of constant magnitude, $|E_b|$, and time-varying voltage angle $\delta_b(t)$, $b \in \mathcal{B}_G$, which is assumed equal to the generator's rotor angle. These voltage sources are located at the internal generator nodes, $\mathcal{B}_G$, and connect to the terminal generator nodes via a reactance $x_b > 0$ (this is the so-called "classical model" for a generator).

Mechanically, a generator consists of a rotor that is driven by a mechanical torque, and whose equations of motion are described by the *swing equation*:

$$\frac{2H_b}{\omega_R}\ddot{\delta}_b + \frac{D_b}{\omega_R}\dot{\delta}_b = P_b - P_b^e, \quad b \in \mathcal{B}_G,$$

where H_b is the machine's inertia constant (in s); D_b is its combined damping coefficient (in s); ω_R is the system's reference frequency (in rad/s); δ_b is the rotor's angle relative to a frame rotating at ω_{sync} (this quantity's expression is given in Sect. 5.2); P_b is the net mechanical power input to the rotor (in p.u.); and P_b^e is the electrical power demanded from the generator via the rest of the network (in p.u.). *Per unit* (p.u.) quantities mean that the powers are normalized by the system's base MVA.[1] A very short time period is considered in power system stability studies. Consequently, it is assumed that P_b and $|E_b|$ are constant for all $b \in \mathcal{B}_G$ and that the transmission lines are lossless (i.e., purely imaginary).

For load nodes it is assumed that power consumption is frequency-dependent and modeled by a first-order differential equation: $P_b^D(t) = P_b^{D\star}(t) + \frac{D_b}{\omega_R}\dot{\delta}_b(t)$, where $D_b > 0$ is a constant and $\delta_b(t) := \phi_b(t) - \omega_{\text{sync}}t$, $b \in \mathcal{B}_L^{\text{new}} \cup \mathcal{B}_L^{\text{term}}$ (recall that $P_b^\star = P_b^{D\star} = 0$, $b \in \mathcal{B}_L^{\text{term}}$). Thus, for loads $\delta_b(t)$ is interpreted as the voltage angle deviation from a reference frame rotating at ω_{sync}, and it is assumed that the actual power demanded, P_b^D, deviates from $P_b^{D\star}$ as a function of $\dot{\delta}_b$.

In general, the active power-flow across a reactance x with voltage sources $|V_b| \exp(i\delta_b)$ and $|V_h| \exp(i\delta_h)$ at either end is given by the *power-angle equation*: $P = \frac{|V_b V_h|}{x}\sin(\delta_b - \delta_h)$. Using this along with the differential equations above we arrive at the equations of motion governing the connected oscillators.

---

[1] An arbitrary value to conveniently scale the quantities of the power grid.

Given a graph of N nodes with power-flow set-point $(P_b^\star, Q_b^\star, P_b^{D\star}, Q_b^{D\star}, |V_b^\star|, \phi_b^\star)$, $b \in \{1, 2, \dots, N\}$, the dynamics of the $\hat{N} := N + N_G$ connected oscillators is described by the following equations:

$$\frac{2H_b}{\omega_R}\ddot{\delta}_b + \frac{D_b}{\omega_R}\dot{\delta}_b = P_b^\star - \left|E_b^\star V_b^\star/x_b\right| \sin\left(\delta_b - \delta_{b+N_G}\right), \quad b \in \mathcal{B}_G,$$

$$\frac{D_b}{\omega_R}\dot{\delta}_b = -\left|E_{b'}^\star V_b^\star/x_b\right| \sin\left(\delta_b - \delta_{b'}\right)$$
$$- \sum_{\substack{h=N_G+1,\\ h\neq b}}^{\hat{N}} \left|V_{b'}^\star V_{h'}^\star Y_{bh}^{\mathrm{SP}}\right| \sin\left(\delta_b - \delta_h\right), \quad b \in \mathcal{B}_L^{\mathrm{term}},$$

$$\frac{D_b}{\omega_R}\dot{\delta}_b = -P_{b'}^{D\star} - \sum_{\substack{h=N_G+1,\\ h\neq b}}^{\hat{N}} \left|V_{b'}^\star V_{h'}^\star Y_{bh}^{\mathrm{SP}}\right| \sin\left(\delta_b - \delta_h\right), \quad b \in \mathcal{B}_L^{\mathrm{new}},$$

where Y_{bh}^{SP} is the element of row b and column h of Y^{SP}; $b' := b - N_G$; $h' := h - N_G$; and the expression for $E_b^\star$ is:

$$\left|E_b^\star\right|^2 = \left(\frac{P_b^\star x_b}{|V_b^\star|}\right)^2 + \left(|V_b^\star| + \frac{Q_b^\star x_b}{|V_b^\star|}\right)^2, \quad b \in \mathcal{B}_G,$$

see [8]. We will use the MATLAB toolbox provided in the paper [8] to compute these modeling constants from a power-flow set-point and physical machine parameters.

In the sequel it will be convenient to consider the dynamics in the following compact form:

$$\frac{2H_b}{\omega_R}\ddot{\delta}_b + \frac{D_b}{\omega_R}\dot{\delta}_b = \hat{P}_b^\star - \sum_{h=1}^{\hat{N}} K_{bh} \sin\left(\delta_b - \delta_h\right), \quad b \in \mathcal{B}_G \cup \mathcal{B}_L^{\mathrm{term}} \cup \mathcal{B}_L^{\mathrm{new}}, \quad (2.14)$$

with $H_b = 0$ for $b \in \mathcal{B}_L^{\mathrm{term}} \cup \mathcal{B}_L^{\mathrm{new}}$; and the matrix $K \in \mathbb{R}^{\hat{N} \times \hat{N}}$ appropriately formed from the constants multiplying the sine terms in the equations above. We will indicate the net active and reactive power generation at a node by:

$$\hat{P}_b := P_b - P_b^D$$

and

$$\hat{Q}_b := Q_b - Q_b^D,$$

respectively. Moreover, we will use $\hat{P}_b^\star := P_b^\star - P_b^{D\star}$ and $\hat{Q}_b^\star := Q_b^\star - Q_b^{D\star}$.

References

1. Bergen AR, Hill DJ (1981) A structure preserving model for power system stability analysis. IEEE Trans Power Appar Syst 1:25–35
2. Bose S, Gayme DF, Low S, Chandy KM (2011) Optimal power-flow over tree networks. In: 2011 49th annual Allerton conference on communication, control, and computing (Allerton), pp 1342–1348
3. Carpentier J (1962) Contribution to the economic dispatch problem. Bull Soc Francoise Electr 3(8):431–447
4. Gan L, Li N, Topcu U, Low S (2012) On the exactness of convex relaxation for optimal power-flow in tree networks. In: 2012 IEEE 51st IEEE conference on decision and control (CDC), pp 465–471
5. Glover JD, Overbye TJ, Sarma MS (2017) Power system analysis and design, 6th edn. Cengage Learning
6. Kocuk B, Dey SS, Sun XA (2016) Inexactness of SDP relaxation and valid inequalities for optimal power-flow. IEEE Trans Power Syst 31(1):642–651
7. Lavaei J, Low SH (2012) Zero duality gap in optimal power flow problem. IEEE Trans Power Syst 27(1):92–107
8. Nishikawa T, Motter AE (2015) Comparative analysis of existing models for power-grid synchronization. New J Phys 17:015012
9. Zimmerman RD, Murillo-Sánchez CE, Thomas RJ (2011) MATPOWER: steady-state operations, planning, and analysis tools for power systems research and education. IEEE Trans Power Syst 26(1):12–19

Chapter 3
Providing Flexibility via Residential Batteries

In this chapter, we discuss how residential batteries within microgrids (MGs) can be used to provide flexibility to the DSO. On this lowest level of the grid hierarchy we only consider active power demand. In particular, we manipulate the aggregated power demand by charging and discharging residential batteries while neglecting the grid topology.

From a DSO perspective one MG can be viewed as one load bus $b \in \mathcal{B}_L$ with flexible active power demand. We collect all MGs in $\mathcal{B}_{MG} \subset \mathcal{B}_L$ and denote the aggregated active power demand of MG b by P_b^D for $b \in \mathcal{B}_{MG}$. In this chapter, we focus on the operation of a single MG $b \in \mathcal{B}_{MG}$. For the sake of notation we drop the dependency of b whenever possible.

3.1 Modelling Microgrids

We use the model introduced in [6] and extended in [3]. The MG consists of $\mathcal{I} \in \mathbb{N}$ households, see also Fig. 3.1. We consider system dynamics

$$x_i(n+1) = \alpha_i x_i(n) + T\left(\beta_i u_i^+(n) + u_i^-(n)\right) \tag{3.1a}$$
$$p_i(n) = u_i^+(n) + \gamma_i u_i^-(n) + w_i(n), \tag{3.1b}$$

where $x_i(n) \in \mathbb{R}$ denotes the State of Charge (SoC) of household i, $i \in \{1, 2, \ldots, \mathcal{I}\}$, at time instant n, $n \in \mathbb{N}_0$, in kWh, $u_i(n) = (u_i^+(n), u_i^-(n))^\top \in \mathbb{R}^2$ the charging/discharging rate in kW, $w_i(n) = \ell_i(n) - g_i(n) \in \mathbb{R}$ the net consumption before battery usage (load minus generation) in kW, and $p_i(n) \in \mathbb{R}$ the resulting power demand in kW at time instant $n \in \mathbb{N}_0$. The dimensionless parameters $\alpha_i, \beta_i, \gamma_i \in (0, 1]$

T. Aschenbruck et al., *Hierarchical Power Systems: Optimal Operation Using Grid Flexibilities*, SpringerBriefs in Energy, https://doi.org/10.1007/978-3-031-25699-8_3

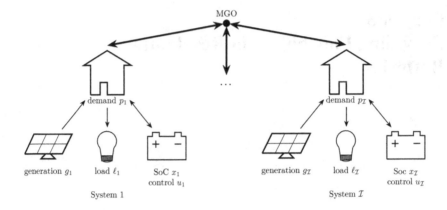

Fig. 3.1 One MG as a network of $\mathcal{I}$ residential energy systems each equipped with generation and storage device. Each subsystem can directly communicate to the microgrid operator (MGO)

describe efficiencies due to self-discharge and energy conversion, while $T > 0$ measures the length of a time step in h. The aggregated power demand within the entire MG is then given by $P_b^D(n) = \sum_{i=1}^{\mathcal{I}} p_i(n)$

Each system's battery is subject to constraints

$$0 \quad \le \quad x_i(n) \quad \le C_i \qquad (3.2\text{a})$$

$$\underline{u}_i \quad \le \quad u_i^-(n) \quad \le 0 \qquad (3.2\text{b})$$

$$0 \quad \le \quad u_i^+(n) \quad \le \overline{u}_i \qquad (3.2\text{c})$$

$$0 \quad \le \quad \frac{u_i^+(n)}{\overline{u}_i} + \frac{u_i^-(n)}{\underline{u}_i} \quad \le 1 \qquad (3.2\text{d})$$

for battery capacity $C_i \ge 0$ and extremal (dis-)charging rates $-\underline{u}_i, \overline{u}_i \ge 0$. We allow both charging and discharging of a single battery within one time step, i.e., we do not enforce $u_i^+(n) \cdot u_i^-(n)$ to vanish. However, we introduce constraint (3.2d) in order to balance these controls. This formulation increases the total amount of flexibility since batteries may dissipate energy if desired. This fact is elaborated when computing the charging rate u_i^u that maximizes the active power demand in Sect. 3.3.

At current time instant $k \in \mathbb{N}_0$ the SoC $x_i(k) = x_i^k$ is measured. The future net consumption is not known in advance. However, we assume that it can be predicted over the next N, $N \in \mathbb{N}_{\ge 2}$, time steps and use the shorthand notation $p_i = (p_i(k), \ldots, p_i(k + N - 1))^\top$.

3.2 Peak Shaving

One challenge that the DSO has to face is the volatile nature of the aggregated power demand profile. It is even amplified due to generation via renewable energy sources, since they typically produce most energy during time of low consumption. Hence, one goal from a DSO's point of view is *peak shaving*, i.e., to achieve a preferably constant power demand and, hence, control energy, which reduces the risk of an outage due to overload. This can be achieved by minimizing the deviation of the aggregated power demand $P_b^D \in \mathbb{R}^N$ from some desired (e.g. constant) reference trajectory $\zeta \in \mathbb{R}^N$ over the prediction horizon, i.e.,

$$\min \quad \left\| P_b^D - \zeta \right\|_2^2.$$

Note that since (3.1b) is linear the aggregated power demand can be written as

$$P_b^D = \sum_{i=1}^{\mathcal{I}} p_i = \sum_{i=1}^{\mathcal{I}} (A_i u_i + w_i)$$

with appropriate matrices $A_i \in \mathbb{R}^{N \times 2N}$. Therefore, the optimization problem can be formulated as

$$\min_u \quad J(u) = \|Au - v\|_2^2 \tag{3.3a}$$

$$\text{s.t.} \quad Du \leq d, \tag{3.3b}$$

with $v = (\zeta - W)$ for $W = \sum_{i=1}^{\mathcal{I}} w_i \in \mathbb{R}^N$. Moreover, the control values are stacked via $u = (u_1^\top, \ldots, u_{\mathcal{I}}^\top)^\top \in \mathbb{R}^{2\mathcal{I}N}$ and A is a block-diagonal matrix with blocks A_i. The matrix D and vector d are chosen such that (3.3b) collects the constraints (3.2) as well as the dynamics (3.1a) for all subsystems. Thus, the initial value x^k is incorporated in $d = d(x^k)$.

There are different approaches to tackle Problem (3.3). A straight-forward option is to solve it in a centralized fashion, i.e., gather all information in one place and solve the optimization problem all at once. This approach comes along with several shortcomings such as the large scale of the optimization problem, no plug-and-play capability as well as data privacy concerns. Instead we use state-of-the-art distributed optimization algorithms such as the alternating direction method of multipliers (ADMM) [2] or the augmented Lagrangian based alternating direction inexact Newton (ALADIN) method [4] to solve (3.3). For implementation details we refer to [3, 5], respectively.

Note that in practice problem (3.3) is solved repeatedly based on updated measurements and forecasts at each time step $k \in \mathbb{N}$. This receding-horizon procedure is referred to as model predictive control (MPC), see e.g. [7] for details. This paper, however, serves as proof that the residential batteries can be used to store energy if the generation on the TSO level (e.g. via renewables) exceeds the demand and set it free otherwise, and help in that way to avoid congestions in the DSO and TSO grids. We, therefore, focus on two time steps. The presented method can be applied repeatedly and, thus, implemented in an MPC framework in a straight-forward manner.

3.3   Generating Flexibility

In order to provide flexibility to the upper grid level, we manipulate the active power
demand $P_b^{D,\text{opt}}(k) = W(k) + A u^{\text{opt}}(k)$ by disturbing the charging behaviour at the
current time instant k yielding

$$P_b^D(k) = W(k) + A u(k) = P_b^{D,\text{opt}}(k) + \Delta.$$

Note that, the current state of charge plays a key role in determining the range of
flexibility the MG can provide. Assume for instance, that all batteries are completely
empty. Clearly, the power demand cannot be reduced by further discharging the
batteries. Similarly, fully charged batteries yield less flexibility.

To determine the maximal possible range for Δ, each household i computes its
extremal charging and discharging rate $\hat{u}_i^+(k)$ and $\hat{u}_i^-(k)$, respectively, depending on
the current SoC $x_i^k = x_i(k)$ and the battery dynamics (3.1a) as well as the constraints
(3.2). In particular, constraint (3.2a) in combination with dynamics (3.1a) yields

$$0 \leq \alpha_i x_i(n) + T\left(\beta_i u_i^+(n) + u_i^-(n)\right) \leq C_i.$$

Therefore, the control that yields the lowest feasible active power demand is given
by

$$\hat{u}_i^-(k) = \left(0 - \min\{\tfrac{\alpha_i}{T} x_i^k, \ -\underline{u}_i\}\right)^\top.$$

Regarding the highest feasible active power demand the computations are more
involved. This is due to the fact that efficiencies $\beta_i, \gamma_i < 1$ ensure that energy dis-
sipates. In particular, if battery i is full, i.e., $x_i(n) = C_i$, then the power demand
might be increased by first discharging and then charging the battery within the same
time step. This effect gains importance if β_i and γ_i are small and batteries can be
discharged faster than charged, i.e., $\overline{u}_i < |\underline{u}_i|$. In order to determine the optimal con-
trol strategy for maximizing the active power demand, each household solves the
small-scale linear programming

$$\hat{u}_i^+(k) = \arg\min \quad -\left(1 \ \gamma_i\right) u_i$$
$$\text{s.t.} \quad D_i^k u_i \leq d_i^k,$$

where D_i^k and d_i^k collect the inequality constraints (3.3b) of system i at current time
k. Since we are only interested in the impact on the power demand, we discount
the discharging rates in terms of efficiencies γ_i. The upper and lower bound on the
aggregated charging rate is then given by

$$U^+(k) = \sum_{i=1}^{\mathcal{I}} \left(1 \; \gamma_i\right) \hat{u}_i^+(k) \quad \text{and} \quad U^-(k) = \sum_{i=1}^{\mathcal{I}} \left(1 \; \gamma_i\right) \hat{u}_i^-(k). \tag{3.4}$$

Thus, from (3.1b) we infer

$$\Delta \in W(k) + [U^-(k), U^+(k)] - P_b^{D,\text{opt}}(k).$$

This is the amount of flexibility the MG can provide. In particular, each $U(k) \in [U^-(k), U^+(k)]$ yields a power demand $P_b^D(k) = W(k) + U(k)$ that deviates from the optimum $P_b^{D,\text{opt}}(k)$ by $\Delta = P_b^D(k) - P_b^{D,\text{opt}}(k)$.

3.4 Associating Costs with the Flexibility

Deviating from the power demand $P_b^{D,\text{opt}}$ is not optimal from the perspective of a microgrid operator (MGO). Therefore, we associate costs with the flexibility. Since we are only interested in structural effects of these costs we choose to penalize the deviation from the optimal value $J(u^{\text{opt}})$.

The flexibility Δ corresponds to fixing the control input u^k at time k such that

$$P_b^D(k) = P_b^{D,\text{opt}}(k) + \Delta = W(k) + a^\top u^{\text{opt}}(k) + \Delta = W(k) + a^\top u^k,$$

where, $a = \left(1 \; \gamma_1 \; \cdots \; 1 \; \gamma_{\mathcal{I}}\right)^\top \in \mathbb{R}^{2\mathcal{I}}$ aggregates all control values at one time instant. Thus, the costs of the flexibility are given by

$$V(\Delta) = \min_{u \in \mathbb{U}(\Delta)} J(u) - J(u^{\text{opt}}) \geq 0, \tag{3.5}$$

where

$$\mathbb{U}(\Delta) = \left\{ u \in \mathbb{R}^{2\mathcal{I}N} \,|\, Du \leq d \text{ and } a^\top u(k) = a^\top u^{\text{opt}}(k) + \Delta \right\}$$

collects all feasible control sequences.

Our simulations in Chap. 7 show that $V(\Delta) \approx \Delta^2$ if the batteries are neither full nor empty at the beginning, i.e., $\Delta \mapsto \Delta^2$ suffices as approximation of the costs of flexibility. Moreover, the resulting costs are easy to communicate since only the optimal aggregated power demand at the current time instant needs to be sent to the DSO.

Note that the coordinated operation of coupled MGs might improve the performance with respect to peak shaving [1, 8]. However, this only affects the costs but not the total amount of flexibility. As this book serves as proof of concept, we treat MGs separately for simplicity.

3.5   Communication to the DSO

The MGs communicate only the information at the current time instant $k \geq 0$ to the DSO, i.e., the flexibility range in the form of lower and upper bound on the active power demand $P_b^{D,-}(k) = W(k) + U^-(k)$ and $P_b^{D,+}(k) = W(k) + U^+(k)$, respectively, and the optimal value $P^{\mathrm{opt}}(k)$. In the future, however, one might incorporate more than one time step on the higher levels of the grid hierarchy, in which case the communication structure would have to be adjusted accordingly.

References

1. Baumann M, Grundel S, Sauerteig P, Worthmann K (2019) Surrogate models in bidirectional optimization of coupled microgrids. Automatisierungstechnik 67(12):1035–1046
2. Boyd S, Parikh N, Chu E, Peleato B, Eckstein J (2011) Distributed optimization and statistical learning via the alternating direction method of multipliers. Found Trends Mach Learn 3(1):1–122
3. Braun P, Faulwasser T, Grüne L, Kellett CM, Weller SR, Worthmann K (2018) Hierarchical distributed ADMM for predictive control with applications in power networks. IFAC J Syst Control 3:10–22
4. Houska B, Frasch JV, Diehl M (2016) An augmented Lagrangian based algorithm for distributed nonconvex optimization. SIAM J Optim 26(2):1101–1127
5. Jiang Y, Sauerteig P, Houska B, Worthmann K (2021) Distributed optimization using ALADIN for model predictive control in smart grids. IEEE Trans Control Syst Technol 29(5):2142–2152
6. Ratnam EL, Weller SR, Kellett CM (2013) An optimization-based approach for assessing the benefits of residential battery storage in conjunction with solar PV, pp 1–8
7. Rawlings JB, Mayne DQ, Diehl M (2017) Model predictive control: theory, computation, and design. Nob Hill Publishing
8. Sauerteig P. Bidirectional optimisation for load shaping within coupled microgrids. Submitted

Chapter 4
Flexibility in the Distribution Grid

In this chapter, we describe how to determine the flexibility of a distribution grid given the flexibility of a set of microgrids (see Chap. 3). Specifically, we compute the minimum, maximum, and optimal active power demand of the entire distribution grid, i.e., the amount of power which must be supplied by the transmission grid. We compute the optimal active power demand of the distribution grid by solving the power-flow equations, and we compute the minimum and maximum active power demands by solving two optimization problems (which contain the power-flow equations as constraints). Furthermore, we present a clustering approach for computing a surrogate model of the distribution grid. It is more computationally efficient to solve the power-flow equations and the optimization problems using the surrogate model than the original full-order model. However, straightforward clustering neglects the complex power losses of lines that are completely contained within a single cluster. We use a heuristic approach to compensate for these losses.

In Sect. 4.1, we describe the computation of the minimum, maximum, and optimal active power demands, and the clustering approach is given in Sect. 4.2. Next, we describe how to compensate for the neglected intracluster line power losses in Sect. 4.3. Finally, we explain which quantities are communicated to the TSO in Sect. 4.4.

4.1 Flexibility Problems

We consider distribution grids which have a single generator (with index b_{ref}) representing the connection to the transmission grid, and the remaining buses are (potentially flexible) loads. In this work, the flexible load buses represent microgrids with flexible real power demand. Consequently, the power needed from the transmission grid is also flexible. Throughout this chapter, we let $\mathcal{B}_L$ represent the load buses on the DSO level, whereas $\mathcal{B}_{MG} \subset \mathcal{B}_L$ represents the microgrid nodes, each consisting of many households and batteries.

© The Author(s), under exclusive license to Springer Nature Switzerland AG 2023
T. Aschenbruck et al., *Hierarchical Power Systems: Optimal Operation Using Grid Flexibilities*, SpringerBriefs in Energy,
https://doi.org/10.1007/978-3-031-25699-8_4

We solve the power-sflow equations (2.2) in order to obtain the active power required from the transmission grid, $P_{b_{ref}}^{opt}$, corresponding to the optimal power demand of the microgrids. Here, $|V_{b_{ref}}|$ and $\phi_{b_{ref}}$ as well as P_b^D and Q_b^D for $b \in \mathcal{B}_L$ are specified. The real power demands of the microgrids, P_b^D for $b \in \mathcal{B}_{MG}$, are the optimal power demands described in Chap. 3. None of the load buses generate power, i.e., $P_b = Q_b = 0$ for $b \in \mathcal{B}_L$.

Next, we determine the minimum and maximum real power needed from the transmission grid, $P_{b_{ref}}^{\ell}$ and $P_{b_{ref}}^u$, by solving the following optimization problem for $\sigma = 1$ and $\sigma = -1$, respectively.

$$\min_{\{|V_b|, \phi_b, P_b, Q_b, P_b^D, Q_b^D\}_{b \in \mathcal{B}}} \sigma P_{b_{ref}}, \tag{4.1a}$$

subject to

$$\phi_{b_{ref}} = \phi_{spec}, \tag{4.1b}$$

$$P_b^{D,-} \le P_b^D \le P_b^{D,+}, \quad b \in \mathcal{B}, \tag{4.1c}$$

$$Q_b^{D,-} \le Q_b^D \le Q_b^{D,+}, \quad b \in \mathcal{B}, \tag{4.1d}$$

$$(2.2),\ (2.5)-(2.7). \tag{4.1e}$$

The objective function in (4.1a) is the real power supplied to the distribution grid by the transmission grid, and the voltage phase angle of the generator is specified in (4.1b). Furthermore, (4.1c) and (4.1d) are respectively bounds on the active and reactive power demand. Finally, (4.1e) represents the power-flow equations, the thermal limits on the absolute value of the complex line currents, and the bounds on the voltage magnitudes and phase angles as well as on the complex power generation in Chap. 2.

None of the loads (including the microgrids) generate power, i.e., $P_b^+ = P_b^- = 0$ for $b \in \mathcal{B}_L$. Furthermore, the real and reactive power demands are specified for non-microgrid loads: $P_b^{D,+} = P_b^{D,-}$ and $Q_b^{D,+} = Q_b^{D,-}$ for $b \in \mathcal{B}_L \backslash \mathcal{B}_{MG}$. Finally, for the microgrid buses, the real and reactive power demand is flexible, i.e., $P_b^{D,+} > P_b^{D,-}$ and $Q_b^{D,+} > Q_b^{D,-}$ for $b \in \mathcal{B}_{MG}$. $P_b^{D,+}$ and $P_b^{D,-}$ are computed using peak-shaving as described in Chap. 3, and in this work, we infer the imaginary parts by assuming a constant ratio between active and reactive power demand for each microgrid.

4.2 Surrogate Model of the Distribution Grid

For large distribution grids, the computational complexity of solving the optimization problem (4.1) can be prohibitively high for real-time application. Therefore, we construct a *surrogate model* of the distribution grid by clustering load buses and solve (4.1) based on the surrogate model.

We briefly illustrate the concept of model reduction (the process of deriving the surrogate model) using a set of nonlinear equations, $f(x) = 0$, where x and $f(x)$ are of the same dimension. The surrogate model is $\hat{f}(\hat{x}) = M_f^T f(M_x \hat{x}) = 0$ where $\hat{x}$ and $\hat{f}(\hat{x})$ have the same dimension which is significantly lower than that of x and $f(x)$. The matrices M_x and M_f are called *projection matrices*, and for the model reduction methods considered in this work, $M_x = M_f$. The projection matrices are chosen such that a solution to $\hat{f}(\hat{x}) = 0$ can be used to approximate that of $f(x) = 0$ by $x \approx M_x \hat{x}$.

4.2.1 The Power-Flow Equations in Vector-Form

Before describing the computation of the projection matrix and the derivation of the surrogate model, we rewrite the power-flow equations (2.2) in vector form for brevity of notation. First, we introduce the matrix $Y^{eq} \in \mathbb{R}^{N \times N}$ where the entries are the *equivalent admittances*,

$$
Y^{eq}_{bh} = \begin{cases} \sum_{h \in \mathcal{N}_S^{(b)}} \left(Y_{bh} + Y_{bh}^{SH}\right) + \sum_{h \in \mathcal{N}_E^{(b)}} \left(Y_{hb} + Y_{hb}^{SH}\right) + \tilde{Y}_b & \text{if } h = b, \\ -Y_{bh} & \text{if } h \in \mathcal{N}_S^{(b)}, \\ -Y_{hb} & \text{if } h \in \mathcal{N}_E^{(b)}, \\ 0 & \text{otherwise,} \end{cases} \quad (4.2)
$$

for $b, h \in \mathcal{B}$. As described in Chap. 2, $h \in \mathcal{N}_S^{(b)}$ means that bus h is connected to bus b through a line *starting* in bus b. Similarly, if $h \in \mathcal{N}_E^{(b)}$, the two buses are connected through a line *ending* in bus b. Next, the power-flow equations can be expressed as

$$
S_b - S_b^D = \sum_{h=1}^{N} \left(Y_{bh}^{eq} V_h\right)^H V_b, \quad b \in \mathcal{B}, \quad (4.3)
$$

or in vector form,

$$
S - S^D = \left(\bar{Y}^{eq} \odot V V^H\right) 1_N. \quad (4.4)
$$

Here, the matrix $\bar{Y}^{eq}$ is the conjugate of Y^{eq}, i.e., $\bar{Y}_{bh}^{eq} = (Y_{bh}^{eq})^H$. Furthermore, all elements of the column vector $1_N \in \mathbb{R}^N$ are ones, $\odot$ denotes the Hadamard product (element-wise multiplication), and for a complex column vector V, V^H is the Hermitian transpose of V, i.e., V^H is a row vector.

Similarly, we write the expressions for the complex line currents (2.4) as

$$I^S = \left(Y + Y^{SH}\right) \odot \left(V1_N^T\right) - Y \odot \left(1_N V^T\right), \tag{4.5a}$$

$$I^E = \left(Y + Y^{SH}\right) \odot \left(1_N V^T\right) - Y \odot \left(V1_N^T\right). \tag{4.5b}$$

If no line connects bus b and h, the corresponding elements of I^S, I^E, Y, and Y^{SH} are zero.

4.2.2 Clustering Based on Proper Orthogonal Decomposition

Proper orthogonal decomposition (POD) [1] is a model reduction method in which the projection matrix is computed using the singular value decomposition (SVD) of a *snapshot matrix* (a matrix whose columns consist of different solutions to the given problem). In contrast, clustering is based on the identification of model variables or quantities with similar behavior, e.g., buses with similar complex voltages and power generation and demand. These variables or quantities are then clustered together. An advantage of clustering is that the resulting surrogate model might retain the structure of the original equations as well as its physical meaning. That is the case for the system considered here. In this work, we choose a clustering projection matrix which approximates the one obtained using POD. Previous work based on this approach (for a dynamical power grid model) showed promising results [2].

The clustering is based on a partitioning of the bus indices $\mathcal{B} = \{1, \dots, N\}$ into n disjoint subsets, B_j for $j = 1, \dots, n$, covering $\mathcal{B}$. In the following, we describe how to obtain this partitioning. First, we solve the optimization problem (4.1) for different upper and lower bounds on the complex power demand for the microgrid buses (and for both $\sigma = 1$ and $\sigma = -1$). In total, we solve M optimization problems. For the kth set of upper and lower bounds, we denote the solution by $V^{(k)} = |V^{(k)}| \odot \exp(i\phi^{(k)})$, $S^{(k)} = P^{(k)} + iQ^{(k)}$, and $S^{D,(k)} = P^{D,(k)} + iQ^{D,(k)}$. For loads that are not microgrids, only the complex voltage is a decision variable. Therefore, we create two snapshot matrices, $X_{|V|}, X_\phi \in \mathbb{R}^{K \times M}$, where $K = |\mathcal{B}_L \backslash \mathcal{B}_{MG}|$ is the number of non-flexible loads and the kth columns contain $|V_b^{(k)}|$ and $\phi_b^{(k)}$ for $b \in \mathcal{B}_L \backslash \mathcal{B}_{MG}$, respectively. Then, as in POD, we compute the *empirical eigenfunctions* $U_{|V|}, U_\phi \in \mathbb{R}^{K \times K}$ by computing the SVDs of the snapshot matrices:

$$X_{|V|} = U_{|V|} \Sigma_{|V|} R_{|V|}^T, \tag{4.6a}$$

$$X_\phi = U_\phi \Sigma_\phi R_\phi^T. \tag{4.6b}$$

Here, $U_{|V|}$ and U_ϕ are orthogonal matrices whose columns consist of the left-singular vectors, $\Sigma_{|V|}, \Sigma_\phi \in \mathbb{R}^{K \times M}$ are rectangular matrices whose diagonal elements are the singular values, and $R_{|V|}, R_\phi \in \mathbb{R}^{M \times M}$ are orthogonal matrices whose columns consist of right-singular vectors (we use nonstandard notation to avoid conflicts with the symbol V for complex voltage). The columns of $X_{|V|}$ and X_ϕ, which correspond to different solutions of the optimization problem, are linear combinations of the columns of $U_{|V|}$ and U_ϕ. Consequently, they can be used as projection matrices. However, as these matrices are square, this would not lead to a reduction in the number of variables. Therefore, we disregard the columns of $U_{|V|}$ and U_ϕ corresponding to singular values smaller than 100ϵ where $\epsilon = 1.1102 \times 10^{-16}$ is the machine precision. As the corresponding singular values are small, we expect the loss in accuracy by disregarding these columns to be small as well. We denote the numbers of columns satisfying this condition by $M_{|V|}^r$ and M_ϕ^r and the resulting matrices by $U_{|V|}^r \in \mathbb{R}^{K \times M_{|V|}^r}$ and $U_\phi^r \in \mathbb{R}^{K \times M_\phi^r}$.

Next, we use an unweighted K-means algorithm [3] to cluster the rows of the matrix $(U_{|V|}^r, U_\phi^r) \in \mathbb{R}^{K \times (M_{|V|}^r + M_\phi^r)}$. The motivation is that each row corresponds to a load bus and similar rows indicate a similar dependence on the microgrid flexibility bounds. We consider the reference generator and each of the microgrid load buses to be separate clusters. The result is the partitioning $\{B_j\}_{j=1}^n$ of the bus indices $\mathcal{B}$ mentioned above, i.e., if $b, h \in B_j$, then bus b and h belong to the jth cluster. Finally, the projection matrix is the characteristic matrix, $M_c \in \mathbb{R}^{N \times n}$, of the partitioning $\{B_j\}_{j=1}^n$ which is given by

$$M_{c,bj} = \begin{cases} 1 & \text{if } b \in B_j, \\ 0 & \text{otherwise.} \end{cases} \tag{4.7}$$

The K-means algorithm does not identify the number of clusters. Therefore, in this book, we select it manually. However, there exist methods to determine this number, e.g., the silhouette method [4]. Furthermore, in future work, we will consider power grids with more microgrids which would then also be clustered together.

The derivation of the surrogate model involves repeated use of the following two identities for arbitrary $X \in \mathbb{C}^{N \times N}$ and $Z \in \mathbb{C}^{n \times n}$ where $n \leq N$.

$$1_N = M_c 1_n, \tag{4.8a}$$

$$M_c^T \left(X \odot \left(M_c Z M_c^T \right) \right) M_c = \left(M_c^T X M_c \right) \odot Z. \tag{4.8b}$$

The first identity holds because each bus belongs to only one cluster. Therefore, each row of M_c only contains one nonzero element (which is 1). The second identity follows from the fact that $(M_c Z M_c^T)_{bh} = Z_{ij}$ for all $b \in B_i$ and $h \in B_j$ and $(M_c^T X M_c)_{ij} = \sum_{b \in B_i, h \in B_j} X_{bh}$.

4.2.3 Clustering of the Power-Flow Equations and Inequalities

In order to derive the surrogate model, we cluster the voltage magnitude and phase angles and the real and reactive power generation and demands. This corresponds to

$$|V| \approx M_c |\hat{V}|, \quad P \approx M_c \hat{P}, \quad P^D \approx M_c \hat{P}^D, \tag{4.9a}$$

$$\phi \approx M_c \hat{\phi}, \quad Q \approx M_c \hat{Q}, \quad Q^D \approx M_c \hat{Q}^D, \tag{4.9b}$$

where $|\hat{V}|$, $\hat{P}$, $\hat{P}^D$, $\hat{\phi}$, $\hat{Q}$, and $\hat{Q}^D$ are the quantities of the clustered grid. Consequently, variables corresponding to buses in the same cluster are identical, e.g., $|V_b| = |V_h| = |\hat{V}_i|$ if $b, h \in B_i$. Based on these approximations, the complex voltage, power generation, and power demands are approximated by

$$V \approx M_c \hat{V}, \quad S \approx M_c \hat{S}, \quad S^D \approx M_c \hat{S}^D, \tag{4.10}$$

where $\hat{V} = |\hat{V}| \odot \exp(i\hat{\phi})$, $\hat{S} = \hat{P} + i\hat{Q}$, and $\hat{S}^D = \hat{P}^D + i\hat{Q}^D$. Here, the exponential function is applied element-wise to its argument.

In order to retain the structure of the original power-flow equations, we formulate the surrogate model in terms of

$$\bar{P} = M_c^T M_c \hat{P}, \quad \bar{P}^D = M_c^T M_c \hat{P}^D, \tag{4.11a}$$

$$\bar{Q} = M_c^T M_c \hat{Q}, \quad \bar{Q}^D = M_c^T M_c \hat{Q}^D, \tag{4.11b}$$

instead of $\hat{P}$, $\hat{P}^D$, $\hat{Q}$, and $\hat{Q}^D$. The corresponding complex power generation and demand are

$$\bar{S} = M_c^T M_c \hat{S}, \quad \bar{S}^D = M_c^T M_c \hat{S}^D. \tag{4.12}$$

The matrix $M_c^T M_c$ is diagonal with positive diagonal entries. Consequently, it is invertible and there is a one-to-one relationship between $\bar{S}$ and $\hat{S}$ and between $\bar{S}^D$ and $\hat{S}^D$.

We cluster the power-flow equations (4.4) by (1) substituting the variable approximations, (2) multiplying by M_c^T from the left, and (3) using the identity (4.8a):

$$M_c^T \left(M_c \hat{S} - M_c \hat{S}^D \right) = M_c^T \left(\bar{Y}^{eq} \odot \left(M_c \hat{V} \hat{V}^H M_c^T \right) \right) M_c 1_n. \tag{4.13}$$

Using the identity (4.8b) and the variables (4.12), we obtain

$$\bar{S} - \bar{S}^D = \left(\hat{\bar{Y}}^{eq} \odot \hat{V} \hat{V}^H \right) 1_n, \tag{4.14}$$

where the clustered complex conjugate of the equivalent admittance matrix is

$$\hat{\bar{Y}}^{eq} = M_c^T \bar{Y}^{eq} M_c. \tag{4.15}$$

Alternatively, $\hat{\bar{Y}}_{bh}^{eq} = (\hat{Y}_{bh}^{eq})^H$ where $\hat{Y}^{eq} = M_c^T Y^{eq} M_c$. The reduced power-flow equations (4.14) are in the same form as the original power-flow equations (4.4).

Next, we introduce the clustered current matrices, $\hat{I}^S = M_c^T I^S M_c$ and $\hat{I}^E = M_c^T I^E M_c$. Again, we use (4.8a) and (4.12):

$$\hat{I}^S = M_c^T \left((Y + Y^{SH}) \odot (M_c \hat{V} 1_n^T M_c^T) - Y \odot (M_c 1_n \hat{V}^T M_c^T) \right) M_c, \quad (4.16a)$$

$$\hat{I}^E = M_c^T \left((Y + Y^{SH}) \odot (M_c 1_n \hat{V}^T M_c^T) - Y \odot (M_c \hat{V} 1_n^T M_c^T) \right) M_c \quad (4.16b)$$

Using (4.8b), we obtain

$$\hat{I}^S = \left(\hat{Y} + \hat{Y}^{SH} \right) \odot \left(\hat{V} 1_n^T \right) - \hat{Y} \odot \left(1_n \hat{V}^T \right), \quad (4.17a)$$

$$\hat{I}^E = \left(\hat{Y} + \hat{Y}^{SH} \right) \odot \left(1_n \hat{V}^T \right) - \hat{Y} \odot \left(\hat{V} 1_n^T \right), \quad (4.17b)$$

where the reduced admittance matrices are

$$\hat{Y} = M_c^T Y M_c, \quad (4.18a)$$

$$\hat{Y}^{SH} = M_c^T Y^{SH} M_c. \quad (4.18b)$$

We reduce the variable bounds in the optimization problem (4.1) by multiplying by M_c^T from the left and substituting the variable approximations (4.9). Consequently, for the variables in the surrogate model, we obtain the bounds

$$|\hat{V}^+| = \left(M_c^T M_c \right)^{-1} M_c^T |V^+|, \quad |\hat{V}^-| = \left(M_c^T M_c \right)^{-1} M_c^T |V^-|, \quad (4.19a)$$

$$\bar{P}^+ = M_c^T P^+, \quad \bar{P}^- = M_c^T P^-, \quad (4.19b)$$

$$\bar{Q}^+ = M_c^T Q^+, \quad \bar{Q}^- = M_c^T Q^-, \quad (4.19c)$$

$$\bar{P}^{D,+} = M_c^T P^{D,+}, \quad \bar{P}^{D,-} = M_c^T P^{D,-}, \quad (4.19d)$$

$$\bar{Q}^{D,+} = M_c^T Q^{D,+}, \quad \bar{Q}^{D,-} = M_c^T Q^{D,-}. \quad (4.19e)$$

Similarly, if we denote by T^+, the matrix whose elements are T_{bh}^+ if a line goes from bus b to bus h and zero otherwise, the clustered upper bounds are

$$\hat{T}^+ = M_c^T T^+ M_c. \quad (4.20)$$

To summarize, the reduced power-flow equations are given by (4.14) which can also be written in the form (2.2), and the reduced complex line currents are given by (4.17). The variables in the surrogate model are $|\hat{V}|$, $\hat{\phi}$, $\bar{P}$, $\bar{Q}$, $\bar{P}^D$, and $\bar{Q}^D$, and the variable bounds in the flexibility optimization problem (4.1) for the surrogate model are given by (4.19). The thermal limits on the lines in the surrogate model are (4.20). Finally, as the reference bus is not clustered with other buses, the objective function in (4.1a) and the equality constraint (4.1b) remain unchanged.

4.3 Compensating for Intracluster Line Power Losses

Using (2.3)–(2.4), we write out the expression for the line power losses:

$$S_{bh}^S + S_{bh}^E = Y_{bh}^H |V_b - V_h|^2 + \left(Y_{bh}^{SH}\right)^H \left(|V_b|^2 + |V_h|^2\right). \tag{4.21}$$

In the clustering, we assume that $V \approx M_c \hat{V}$ which implies that the voltages of buses in the same cluster are identical, i.e., $V_h = V_b$ for $b, h \in B_j$. Consequently, the first term in the right-hand side of (4.21) is zero. This creates a discrepancy between the total power loss in the original model and in the surrogate model. We use the training data (originally used to identify the clusters) to compensate for this discrepancy. Furthermore, it is desirable to *overestimate* the minimum power, $P_{b_{ref}}^\ell$, and *underestimate* the maximum power, $P_{b_{ref}}^u$, required from the transmission grid in order to ensure that only feasible bounds are communicated to the TSO. Let

$$\Delta S_j = \sum_{\substack{b,h \in B_j \\ b \neq h}} Y_{bh}^H |V_b - V_h|^2 \tag{4.22}$$

be the complex line power losses in cluster j which are not accounted for in the surrogate model, and let $\Delta S_j^{(k)}$ be the corresponding value for the kth set of training data. Then, we add the real and imaginary part of $\frac{1}{M} \sum_{k=1}^M \Delta S_j^{(k)}$ to the active and reactive power demand of the jth cluster when we compute $P_{b_{ref}}^{opt}$. Similarly, we add $\max_k \text{Re}\,\Delta S_j^{(k)}$ and $\max_k \text{Im}\,\Delta S_j^{(k)}$ to the real and reactive power demand of cluster j when computing $P_{b_{ref}}^\ell$, and we add $\min_k \text{Re}\,\Delta S_j^{(k)}$ and $\min_k \text{Im}\,\Delta S_j^{(k)}$ when computing $P_{b_{ref}}^u$. (Note that we use the maximum values when computing the lower bound and vice versa to obtain conservative approximations.)

4.4 Communication to the TSO Level

We communicate the total active power demand of the distribution grid, $P_{b_{ref}}^{opt}$, corresponding to the optimal microgrid active power demands to the TSO. This is used as the desired power demand in the second term of the objective function (5.1) described in the following chapter. Furthermore, we communicate the minimum, $P_{b_{ref}}^\ell$, and maximum, $P_{b_{ref}}^u$, active power demand of the distribution grid which constitute the bounds in (5.2). Consequently, the TSO knows the range of active powers which can be achieved by manipulating the power demands of the microgrids, i.e., by manipulating the household batteries.

References

1. Antoulas AC (2005) Approximation of large-scale dynamical systems, vol 6. In: Advances in design and control. SIAM Publications, Philadelphia, PA
2. Aschenbruck T, Baumann M, Esterhuizen W, Filipecki B, Grundel S, Helmberg C, Ritschel TKS, Sauerteig P, Streif S, Worthmann K (2021) Optimization and stabilization of hierarchical electrical networks. In: Mathematical modeling, simulation and optimization for power engineering and management, vol 34. Mathematics in industry. Springer, pp 171–198
3. Bishop CM (2006) Pattern recognition and machine learning. In: Information science and statistics. Springer
4. Rousseeuw PJ (1987) Silhouettes: a graphical aid to the interpretation and validation of cluster analysis. J Comput Appl Math 20:53–65

Chapter 5
Security and Stability on the Transmission Grid

In this chapter, we discuss how flexibilities from microgrids conveyed through the distribution level can be utilized in the operation of transmission systems. To this end, we consider both nonlinear (2.2)–(2.7) and semidefinite (2.8)–(2.13) OPF formulations and introduce flexible demand nodes (representing the connections to the DSO level) to analyze the resulting changes in energy consumption and line loading.

We consider two separate problems on this level. The first is the *security-constrained OPF problem*, where we want to ensure that a single line failure will not cause adverse effects over the entire network (the so-called $n - 1$ security). The second problem is to find a power-flow set-point that corresponds to *synchronized generators*.

Throughout this chapter we let $\mathcal{B}_G$ and $\mathcal{B}_L$ represent the generators and loads on the TSO grid, respectively, and let $\mathcal{B}_L^{\text{flex}} \subset \mathcal{B}_L$ denote the indices of the flexible loads, representing the connections to the DSO level. Thus, in general, there may be many independent DSO grids connecting to the TSO grid at multiple distinct nodes (though in the example chapter we will only consider one DSO grid connected to one TSO grid). With each flexible load we associate a triple $(P_b^{D,-}, P_b^{D,+}, P_b^{D,opt})$ representing the minimal, maximal and optimal power demand, respectively, at flexible load $b \in \mathcal{B}_L^{\text{flex}}$. These values are equal to the corresponding values $P_{b_{\text{ref}}}^{\ell}, P_{b_{\text{ref}}}^{u}$ and $P_{b_{\text{ref}}}^{\text{opt}}$ produced from the optimization problems solved on the DSO level.

The cost function of the OPF problems considered in this chapter is then formed as follows,

$$J\left(P, P^D\right) := \sum_{b \in \mathcal{B}_G} \left(c_{b,2} P_b^2 + c_{b,1} P_b + c_{b,3}\right) + \sum_{b \in \mathcal{B}_L^{\text{flex}}} \alpha_b \left(P_b^D - P_b^{D,opt}\right)^2, \quad (5.1)$$

where $c_{b,2}$ is a quadratic cost coefficient, $c_{b,1}$ is a linear cost coefficient and $c_{b,3}$ a constant; $P := (P_1, P_2, \ldots, P_N)^\top$, $P^D := (P_1^D, P_2^D, \ldots, P_N^D)^\top$; and the α_b's are user-specified weights. Thus, we penalize the active power generated at generator

© The Author(s), under exclusive license to Springer Nature Switzerland AG 2023
T. Aschenbruck et al., *Hierarchical Power Systems: Optimal Operation Using Grid Flexibilities*, SpringerBriefs in Energy,
https://doi.org/10.1007/978-3-031-25699-8_5

nodes, as well as the deviation of the power demanded at flexible nodes from the optimals, determined from the optimization problems on the DSO level. The quantities $P_b^{D,-}$ and $P_b^{D,+}$ will appear as constraints in the OPF problem:

$$P_b^{D,-} \leq P_b^D \leq P_b^{D,+}, \quad b \in \mathcal{B}_L^{\text{flex}}. \tag{5.2}$$

5.1 Security-Constrained Optimal Power-Flow

Security analysis has long been a part of problems associated with power-flow [8]. In this book, we consider the $n - 1$ security concept, in which a failure of a single line will not cause other lines to overload, trip and result in a blackout. Note, however, that in an immediate post-contingency scenario, slight violations of line loading limits are usually allowed. While some authors have made investigations into a possibility of power plant failures, we focus here solely on failures of power lines. The security notion can be extended to cover more than one simultaneous failure, which is the case for $n - 2$ security. However, even the $n - 1$ case can be computationally intractable.

There are two possible approaches to ensure security of the network, to either take preventive or curative actions. In the first case, we try to find a setpoint, which will fulfill network requirements for both the standard scenario and for each contingency that can occur, without changing power generation. In the curative approach, we allow the operator to take actions after a contingency occurs, so the network state can be further adjusted. In this book, we focus solely on the preventive approach. We use subscript 0 to denote variables and constraints of the original (non-contingency) scenario and subscript $k > 0$ to denote a contingency associated with failure of line k. We reformulate our problem with objective (5.1) and constraints (2.8)–(2.13), (5.2) as a general-form semidefinite program:

$$\min_{W_0, S_0, S_0^D} J\left(P_0, P_0^D\right) \tag{5.3a}$$

subject to

$$h_0(W_0, S_0, S_0^D) = 0 \tag{5.3b}$$

$$g_0(W_0) \leq 0 \tag{5.3c}$$

$$P_0^- \leq P_0 \leq P_0^+, \tag{5.3d}$$

$$Q_0^- \leq Q_0 \leq Q_0^+, \tag{5.3e}$$

$$P_0^{D,-} \leq P_0^D \leq P_0^{D,+}, \tag{5.3f}$$

$$W^- \leq W_0 \leq W^+ \tag{5.3g}$$

$$W_0 \succeq 0, \tag{5.3h}$$

where our decision variables are W_0, matrix corresponding to the voltage products, $S_0 = P_0 + iQ_0$, complex power generation and $S_0^D = P_0^D + iQ_0^D$, complex power demand. The cost in (5.3a) is the one specified in (5.1). The constraints in (5.3b) are the power balance constraints (2.9); (5.3c) are the constraints on thermal limits (2.10), on both ends of each line. Moreover, the constraints (5.3d)–(5.3f) restrict power generation (2.12), flexible demand and voltage limits (2.11) respectively (these inequalities are interpreted element-wise); and (5.3g) are the voltage limits. Furthermore, we add additional variables W_k and similar constraints corresponding to each contingency:

$$h_k(W_k, S_0, S_0^D) = 0 \quad \forall k \tag{5.4a}$$

$$g_k(W_k) \leq 0 \quad \forall k \tag{5.4b}$$

$$W^- \leq W_k \leq W^+ \quad \forall k \tag{5.4c}$$

$$W_k \succeq 0 \quad \forall k. \tag{5.4d}$$

Here, (5.4a) represents the bus balance equations for each contingency k, after it happens. I.e., this is power balance with a single line missing. Equation (5.4b) are then inequality constraints, including the thermal limits (2.10) for the remaining lines. Note that the equality constraints use S_0, power generation for the original scenario.

In the most basic security-constrained approach, one considers all possible lines as potential points of failure. However, depending on the solution approach, this might quickly make the problem intractable. Power grids can easily have up to thousands of nodes resulting in an extreme number of variables and constraints. While it is possible to solve linear relaxations of this size, the more accurate SDP, second-order cone programming and nonlinear approaches fall short of achieving this kind of computational efficiency. Therefore, we seek a way to reduce the number of contingencies and add only a limited number of constraints to the problem. In this book, we only disregard all lines, whose removal would cause islanding of the network (i.e. would turn it into two disjointed parts). In further research, we will choose contingencies that result in overloadings/violations of security limits.

While disregarding some lines in the contingency constraints increases the computational efficiency, we still have to solve an expensive optimization problem. This is especially the case for semidefinite programming problems since long computation times for larger matrices can quickly make the problem intractable. Alleviating this issue requires further study of decomposition techniques for SDP relaxation of the OPF, which is outside the scope of this book. After a secure solution is obtained we communicate the active power demand of the flexible loads back to the DSO level, where it is propagated back to the microgrids.

5.2 Stability-Guaranteed Power-Flow

A solution to the power-flow equations on the TSO level describes a *static* distribution of power in the network. In this section we consider the *dynamic* problem of the synchronization of connected machines. As we will elaborate, a power-flow set-point may or may not correspond to the existence of a stable equilibrium point for the SP model (2.14). Because it is difficult to enforce state-of-the-art conditions that guarantee synchronism into power-flow optimization problems, in this section we propose a heuristic iterative procedure.

Intuitively, a network of oscillators is synchronized if all individual oscillators rotate at some constant angular speed. Synchronization in complex networks has been the focus of much research, see for example [1–4] for a survey. The focus of these works is the derivation of conditions that imply the existence of synchronized operating points (which correspond to stable equilibria).

Research has also been conducted into how network parameters may be perturbed to improve stability in power networks. The paper [6] considers the SP model and shows, via the implicit function theorem, how damping in the network can be improved through perturbations of the active powers, line impedances and network topology. In [7] stability conditions, akin to the so-called master stability formalism [1], are derived from the linearized dynamics of the so-called *effective network model*. (In this model loads are assumed to be constant impedances, unlike in the SP model covered in Sect. 2.3.) They then show how generator parameters can be tuned to enhance the network's stability. The paper [5] addresses the problem of maximizing the network's so-called *state algebraic connectivity* (the second largest eigenvalue of the Laplacian of the linearized dynamics about an equilibrium point) via optimization problems.

These papers provide valuable insight into the relationship between modeling parameters and stability, but do not consider how power-flow fits into the study. Recall from Sect. 2.3 that the entries of the matrix K appearing in (2.14) are a function of a power-flow set-point $(P_b^\star, Q_b^\star, P_b^{D\star}, Q_b^{D\star}, |V_b^\star|, \phi_b^\star)$, $b \in \{1, 2, \ldots, N\}$.

5.2.1 Sufficient Conditions for Synchronization

The paper [4] presents easily checkable sufficient conditions, involving the system parameters, for the existence of a stable equilibrium point of the SP model (2.14). We first present a number of concepts, as covered in the supporting information of the paper [4]. Staying with the bus indices introduced in Sect. 2.3, let $\mathbb{T}^{\hat{N}} :=$ $\mathbb{S}^1 \times \cdots \times \mathbb{S}^1$ denote the $\hat{N}$-dimensional torus, that is, the product space of $\hat{N}$ circles. (Recall from Sect. 2.3 that the graph with N nodes is augmented to a graph of $\hat{N}$ nodes.) Let $\Delta(\gamma) \subset \mathbb{T}^{\hat{N}}$ be the set of angles $(\delta_1, \ldots, \delta_{\hat{N}}) \in \mathbb{R}^{\hat{N}}$ with the property that $|\delta_b - \delta_h| \leq \gamma$, for $\{b, h\} \in \mathcal{E}$, where $\mathcal{E}$ is the set of edges of the $\hat{N}$-node graph. If $\delta \in \Delta(\gamma)$, we will say that the phase angles are *cohesive* with angle γ.

Definition 5.1 A solution $(\delta, \dot{\delta}) : \mathbb{R}_{\geq 0} \to (\mathbb{T}^{\hat{N}}, \mathbb{R}^{N_G})$ to the coupled oscillator model (2.14) is said to be *synchronized* if $\delta(0) \in \Delta(\gamma)$ and there exists an $\omega_{\text{sync}} \in \mathbb{R}$ such that $\delta(t) = \delta(0) + \omega_{\text{sync}} 1_{\hat{N}} t \pmod{2\pi}$ and $\dot{\delta}(t) \equiv \omega_{\text{sync}} 1_{N_G}$ for all $t \geq 0$, where $1_M \in \mathbb{R}^M$ denotes an M-dimensional vector with all elements equal to 1. Thus, a solution of the connected oscillator model is synchronized if all machines are rotating at the same synchronous frequency and the phases are cohesive with some angle γ. This cohesiveness of bus angles is very important in power grid stabilization studies, and is required to be small (less than $15°$). The SP model's synchronous frequency is explicitly given by $\omega_{\text{sync}} := \sum_{b=1}^{\hat{N}} \hat{P}_b^\star / \sum_{b=1}^{\hat{N}} D_b$ (see the supporting information from [4]), where $\hat{P}_b^\star := P_b^\star - P_b^{D\star}$, see Sect. 2.3. Thus, after substituting $\dot{\delta}_b$ with $\dot{\delta}_b - \omega_{\text{sync}}$, and $\hat{P}_b^\star$ with $\hat{P}_b^\star - D_b \omega_{\text{sync}}$, we can focus on equilibrium points $(\delta^\star, 0) \in \mathbb{T}^{\hat{N}} \times \mathbb{R}^{|N_G|}$ without loss of generality.

The following result presents sufficient conditions for the existence of a unique and stable equilibrium point of the model (2.14).

Proposition 5.1 (See [4]) *The coupled oscillator model (2.14) has a unique and stable equilibrium point* $(\delta^\star, 0) \in \mathbb{T}^{\hat{N}} \times \mathbb{R}^{N_G}$ *with cohesive phases* $|\delta_b^\star - \delta_h^\star| \leq \gamma < \frac{\pi}{2}$ *for every pair of connected oscillators* $\{b, h\} \in \mathcal{E}$ *if*

$$\|L^\dagger \hat{P}^\star\|_{\mathcal{E},\infty} \leq \sin(\gamma). \tag{5.5}$$

Here, $L^\dagger \in \mathbb{R}^{\hat{N} \times \hat{N}}$ is the Moore–Penrose inverse of the network Laplacian matrix,

$$L := \text{diag}\left(\left\{\sum_{j=1}^{\hat{N}} K_{ij}\right\}_{i=1}^{\hat{N}}\right) - K, \tag{5.6}$$

the vector $\hat{P}^\star := (\hat{P}_1^\star, \hat{P}_2^\star, \ldots, \hat{P}_{\hat{N}}^\star)^\top \in \mathbb{R}^{\hat{N}}$ is formed from the constants $\hat{P}_b^\star$ from (2.14), and $\|x\|_{\mathcal{E},\infty} = \max_{\{b,h\} \in \mathcal{E}} |x_b - x_h|$ is the *worst-case dissimilarity* for the vector $x = (x_1, \ldots, x_{\hat{N}})$ over the edges $\mathcal{E}$. As reported in [4], the sufficient condition in (5.5) only fails to predict stability for special degenerate networks, which are very unlikely to appear for real scenarios.

5.2.2 The Stability Algorithm

An attractive aspect of the sufficient condition in (5.5) is that the machines' inertia and damping parameters are not needed: only their transient reactances influence the matrix K, and therefore the Laplacian L. Ideally, one would like to incorporate the inequality (5.5) in the power-flow solver to guarantee a stable operating point for the coupled oscillators at steady-state. However, it is very difficult to analytically express the norm $\|L^\dagger \hat{P}^\star\|_{\mathcal{E},\infty}$ as a function of an arbitrary power-flow set-point.

Moreover, even if this were done, the mapping would be nonlinear and inconvenient to be included in an optimization problem. Consequently, in this section we propose an iterative algorithm that attempts to arrive at a suitable power-flow set-point.

To lighten our notation, let $Z := (\hat{P}_1, \hat{Q}_1, |V_1|, \phi_1, \ldots, \hat{P}_N, \hat{Q}_N, |V_N|, \phi_N) \in \mathbb{R}^{4N}$ denote an arbitrary element of $\mathbb{R}^{4N}$, and let $Z^\star := (\hat{P}_1^\star, \hat{Q}_1^\star, |V_1^\star|, \phi_1^\star, \ldots, P_N^\star, Q_N^\star, |V_N^\star|, \phi_N^\star) \in \mathbb{R}^{4N}$ denote a power-flow set-point, that is, a vector that satisfies the power-flow equations, (2.2) on the N node TSO grid.

There are a few steps from an arbitrary power-flow set-point $Z^\star \in \mathbb{R}^{4N}$ to the pseudo-inverse, $L^\dagger \in \mathbb{R}^{\hat{N} \times \hat{N}}$ (that is, the mapping $Z^\star \mapsto L^\dagger(Z^\star)$). First, compute the matrix K in the dynamical system (2.14). Second, form the Laplacian, L, as in (5.6). Third, compute the inverse, $L^\dagger$. In the sequel we will sometimes explicitly indicate the dependence of the pseudo-inverse $L^\dagger$ and the matrix K on an arbitrary power-flow set-point $Z^\star$ by specifying the argument: $L^\dagger(Z^\star)$ and $K(Z^\star)$. Moreover, we will indicate $L^\dagger$'s ith row by $L^\dagger_{i,\bullet}$.

The procedure, see Algorithm 1, iteratively invokes the following SDP problem:

$$\min_{(Z,\gamma) \in \mathbb{R}^{4N+1}} \quad J(P, P^D) + \alpha_\gamma \gamma, \tag{5.7a}$$

subject to

$$(2.8)-(2.13) \tag{5.7b}$$

$$\left| \left[L^\dagger_{i,\bullet}(Z^\star_{\text{prev}}) - L^\dagger_{j,\bullet}(Z^\star_{\text{prev}}) \right] \hat{P} \right| \leq \gamma, \quad \forall (i,j) \in \mathcal{E}, \tag{5.7c}$$

where $Z^\star_{\text{prev}}$ is a (previously known) solution to (5.7); $\hat{P} := (\hat{P}_1, \hat{P}_2, \ldots, \hat{P}_{\hat{N}})^\top \in \mathbb{R}^{\hat{N}}$; J is the cost function labeled (5.1); and $\alpha_\gamma \in \mathbb{R}_{\geq 0}$ is a user-defined constant that weights the importance of γ in relation to J.

Remark 5.1 In the problem (5.7) the vector $\hat{P} \in \mathbb{R}^{\hat{N}}$ appearing in the inequalities labeled (5.7c) is formed from the vector $Z \in \mathbb{R}^{4N}$ and has zero entries for all elements P_b, with $b \in \mathcal{B}_L^{\text{term}}$ (recall that these indices refer to artificial terminal loads that draw no power, see Sect. 2.3). Thus, the elements of $\hat{P}$ are decision variables contained in Z.

We impose the following assumptions:

(A1) There exists a known initial power-flow set-point $Z_0^\star$ such that $Z = Z_{\text{prev}}^\star = Z_0^\star$ (along with some $\gamma_0 \in \mathbb{R}_{\geq 0}$ which always exists) solves problem (5.7).
(A2) The constant γ is sufficiently small, such that $\sin(\gamma) \approx \gamma$.
(A3) The mapping $Z^\star \mapsto L^\dagger(Z^\star)$ is locally Lipschitz continuous.

Assumption (A1) is needed so that our algorithm is initially feasible. Assumption (A2) allows us to include the linear constraints (5.7c) as a close approximation to (5.5), and to easily penalize γ with a soft constraint in (5.7a). Assumption (A3), the most critical assumption, is needed to so that $L^\dagger$ remains "slow changing" with respect to power-flow set-points $Z^\star$ over a neighborhood of the previous set-point

Z^*_{prev}. This allows us to consider the linear constraints (5.7c) (where $L^\dagger$ is assumed constant) in a tractable optimization problem. The algorithm iteratively invokes the optimization problem (5.7), attempting to find a new power-flow set-point that satisfies the synchronization condition (5.5).

Algorithm 1 Find a power-flow set-point $\bar{Z}^\star \in \mathbb{R}^{4N}$ corresponding to a stable equilibrium point of the $\hat{N}$ oscillators in (2.14)

Require: Initial power-flow set-point $Z^*_0 \in \mathbb{R}^{4N}$; desired cohesiveness $\bar{\gamma} \in \mathbb{R}_{\geq 0}$; max number of iteration $k^{\max} \in \mathbb{N}$.
1: **for** $k \in \{0, 1, \ldots, k^{\max}\}$ **do**
2:    Compute $L^\dagger(Z^*_k)$.
3:    Form $\hat{P}^k \in \mathbb{R}^{\hat{N}}$ from $Z^*_k \in \mathbb{R}^{4N}$, see Remark 5.1.
4:    **if** $\|L^\dagger(Z^*_k)\hat{P}^k\|_{\mathcal{E},\infty} \leq \sin(\bar{\gamma})$ **then**
5:       Let $\bar{Z}^\star \leftarrow Z^*_k$.
6:       **Exit with SUCCESS.**
7:    **else**
8:       Solve optimization problem (5.7) with $Z^*_{\text{prev}} \leftarrow Z^*_k$, to produce $Z^*_{\text{new}}, \gamma_{\text{new}}$
9:       Let $Z^*_{k+1} \leftarrow Z^*_{\text{new}}$.
10:   **end if**
11: **end for**
12: Let $\bar{Z}^\star \leftarrow$ **NULL**.
13: **Exit with NO CONCLUSION.**

5.2.3 Remarks and Perspectives

Note that the stability check says whether a stable operating point corresponding to a power-flow set-point exists, not whether this point can actually be reached from a previous set-point. This is a problem that could be addressed in the future. Assumption (A3) is the most critical one made in this section. In our numerical example in Chap. 7 we observed that the change in $L^\dagger$'s eigenvalues were small across iterations of k in Algorithm 1, indicating that $L^\dagger$ remained slow-changing. In the future we intend to verify, with a large numerical study involving typically-sized TSO grids, whether Assumption (A3) is a reasonable one in general, and whether it holds with small Lipschitz constant. Finally, there are no guarantees that the algorithm will converge to a stable set-point hence the maximum number of iterations, $k^{\max}$. We intend to investigate convergence in the future.

References

1. Arenas A, Díaz-Guilera A, Kurths J, Moreno Y, Zhou C (2008) Synchronization in complex networks. Phys Rep 469(3):93–153
2. Boccaletti S, Latora V, Moreno Y, Chavez M, Hwang D-U (2006) Complex networks: structure and dynamics. Phys Rep 424(4–5):175–308
3. Dörfler F, Bullo F (2014) Synchronization in complex networks of phase oscillators: a survey. Automatica 50(6):1539–1564
4. Dörfler F, Chertkov M, Bullo F (2013) Synchronization in complex oscillator networks and smart grids. Proc Natl Acad Sci 110(6):2005–2010
5. Li B, Michael Wong KY (2017) Optimizing synchronization stability of the Kuramoto model in complex networks and power grids. Phys Rev E 95(1):012207
6. Mallada E, Tang A (2011) Improving damping of power networks: power scheduling and impedance adaptation. In: 2011 50th IEEE conference on decision and control and European control conference. IEEE, pp 7729–7734
7. Motter AE, Myers SA, Anghel M, Nishikawa T (2013) Spontaneous synchrony in power-grid networks. Nat Phys 9(3):191–197
8. Stott B, Alsac O, Monticelli AJ (1987) Security analysis and optimization. Proc IEEE 75(12):1623–1644

Chapter 6
Implementation in the Distribution Grid and the Microgrids

In this chapter, we describe the implementation of the optimal amount of power delivered by the transmission grid in the distribution grid and the microgrids.

6.1 Implementation at the Distribution Grid Level

Once the security-constrained OPF problem from Sect. 5.1 or the stability check from Sect. 5.2 has been solved, the amount of power allocated to the distribution grid is known. However, the amounts of power allocated to the microgrids remain to be determined. We compute these by solving the optimization problem (6.1) which is similar to (4.1) that was solved in order to obtain the minimum and maximum active power demands of the distribution grid in Chap. 4. The two differences are that (1) we minimize the sum of squared deviations from the optimal active microgrid power demands, P_b^{opt} for $b \in \mathcal{B}_{MG}$, and (2) the real power generation of the reference generator is specified in (6.1b) where P_{spec} is the active power supplied by the transmission grid. The remaining constraints (6.1c), are the same as in (4.1). The optimization problem is

$$\min_{\{|V_b|, \phi_b, P_b, Q_b, P_b^D, Q_b^D\}_{b \in \mathcal{B}}} \sum_{b \in \mathcal{B}_{MG}} (P_b^D - P_b^{\text{opt}})^2, \tag{6.1a}$$

subject to

$$P_{b_{\text{ref}}} = P_{\text{spec}}, \tag{6.1b}$$

$$(4.1\text{b})-(4.1\text{e}). \tag{6.1c}$$

The optimal active power demands of the microgrids, P_b^D for $b \in \mathcal{B}_{MG}$, obtained by solving (6.1) are communicated to the microgrid level.

© The Author(s), under exclusive license to Springer Nature Switzerland AG 2023   41
T. Aschenbruck et al., *Hierarchical Power Systems: Optimal Operation Using Grid Flexibilities*, SpringerBriefs in Energy,
https://doi.org/10.1007/978-3-031-25699-8_6

For completeness, in the numerical example presented in Chap. 7, we explore the consequences of using a surrogate model when solving (6.1). We use the same surrogate model as was used to compute $P_{b_{ref}}^{opt}$ in Chap. 4, i.e., we add the average real and reactive line power losses from the training data to the real and reactive power demands of the corresponding clusters.

6.2   Implementation at Microgrids

In this section, we briefly discuss how the desired power demand P_b^D, $b \in \mathcal{B}_{MG}$, from a DSO/TSO's perspective is implemented at the MGs. Similar to Chap. 3, we consider a single MG and drop the index b. Note that if $P^D = W(k)$, then there is no need to interfere at all. Otherwise, we distribute the difference $P^D - W(k)$ equally to all households taking their charging ability into account, i.e., the control of household i can be computed as

$$u_i(k) = \begin{cases} \frac{P^D - W(k)}{P^u(k) - W(k)} u_i^u(k) & \text{if } P^D > W(k) \\ \frac{P^D - W(k)}{P^l(k) - W(k)} u_i^l(k) & \text{if } P^D < W(k). \end{cases}$$

Note that this is always feasible by choice. Based on the control inputs the SoCs are updated according to (3.1a) and the loop is completed.

Chapter 7
Numerical Example

In this chapter, we present a numerical example of the approach presented in the previous chapters based on modifications of small-scale standard test systems. We emphasise that the sizes of these grids are very small and the power demands very low when compared to real-world grids. For example, active power transmission demand in real-world TSO grids can be in the 1000's of MW. Nevertheless, the example presents a small-scale proof-of-concept, and larger networks will be studied in the future.

We consider a modified IEEE 9-bus system as transmission grid where one load bus (node 5) represents an active distribution grid. The single distribution grid is modeled as a modified IEEE 33-bus system containing three flexible microgrid load nodes. We consider two scenarios in the TSO grid:

(A) power generation is high compared to power demand (e.g. due to renewables),
(B) power generation is low compared to power demand.

Scenario (A) corresponds to the first time interval, $[k, k + 1)$ where $k \geq 0$, and scenario (B) corresponds to the following time interval, $[k + 1, k + 2)$. The duration of each time interval is $T = 0.5$ h. We demonstrate that energy can be stored in residential batteries in the microgrids during $[k, k + 1)$ and released again during $[k + 1, k + 2)$, thus, improving the flexibility of the transmission grid operation. Note that, in general, the energy stored in the batteries could be released at any point of time in the future.

7.1 Flexibility of the MGs

In this section, we give a detailed description of the data we used to model the microgrids and discuss the simulation results.

© The Author(s), under exclusive license to Springer Nature Switzerland AG 2023
T. Aschenbruck et al., *Hierarchical Power Systems: Optimal Operation Using Grid Flexibilities*, SpringerBriefs in Energy,
https://doi.org/10.1007/978-3-031-25699-8_7

Table 7.1 MG data used in simulations

α_i	$\mathcal{N}(0.99, 0.01)$	C_i	$\mathcal{N}(2, 0.5)$	kWh	x_i^k	$\mathcal{N}(0.5, 0.05)$	kWh
β_i	$\mathcal{N}(0.95, 0.05)$	$\overline{u}_i$	$\mathcal{N}(0.5, 0.05)$	kW	T	0.5	h
γ_i	$\mathcal{N}(0.95, 0.05)$	$\underline{u}_i$	$\mathcal{N}(-0.5, 0.05)$	kW	N	48	

The parameters α_i, β_i, γ_i, and N are dimensionless variables

The data used in the simulations is listed in Table 7.1, where $\mathcal{N}(\mu, \sigma)$ denotes the normal distribution with mean value μ and standard deviation σ. All values are projected to feasible intervals, e.g., $\alpha_i \in (0, 1]$. The battery efficiencies are based on [7]. For instance, Meinecke et al. assume a self-discharge of 2% per day which refers to $\alpha_i = 0.98^{\frac{1}{48}} \approx 0.9996$. Furthermore, both load and generation profiles of the households (and, hence, $w_i = \ell_i - g_i$) are taken from an Australian grid operator [8]. Since we consider the two scenarios consecutively, we initialize the batteries in (B) with the terminal SoC in (A). Throughout our simulations we choose the reference trajectory ζ to be the aggregated net consumption W averaged over the previous N time steps (the last 24 h), i.e.,

$$\zeta(n) = \frac{1}{N} \sum_{j=n-N+1}^{n} W(j),$$

where we assume the net consumption of the previous N steps, i.e., $w_i(n)$ for all $n \in \{k - N + 1, \ldots, k\}$ and $i \in \{1, \ldots, \mathcal{I}\}$, to be given.

Results of the peak-shaving OCP (3.3) and the impact of providing flexibility within one MG can be found in Fig. 7.1. If not mentioned otherwise all results in this subsection refer to MG 1. Power demand $P_b^{D,\mathrm{opt}}$ and state x^{opt} associated with the optimal solution u^{opt} of (3.3) are represented by the solid black lines (left). The dotted blue line in Fig. 7.1 (left) represents the optimal trajectory after discharging the batteries during the first step while the dash-dot red line stands for charging them. For the sake of readability we only indicate the demand at time $k = 0$; these values can be found in Table 7.3. After the manipulation Δ at $n = k$, the SoC (dotted blue/dash-dot red line bottom) approaches the optimal SoC trajectory (solid black line). The terminal SoC is $x = 0$ independent of the initial disturbance. As a result, the power demand after additional discharge at the beginning (dotted blue line top, left) stays slightly above the optimal demand (solid black line) and the demand after charging (dash-dot red line) stays below.

A visualization of the cost function can be found in Fig. 7.1 (right, solid blue line). The vertical dotted black line marks the net consumption, i.e., the power demand if the batteries are not used. The optimal control associated with peak shaving is to discharge the batteries, see also the decrease of the solid black line in the beginning in Fig. 7.1 (left, bottom). As a consequence, the optimum in Fig. 7.1 (right) is below $W(k)$. By construction of the costs of flexibility in (3.5), discharging is cheaper than charging in that situation.

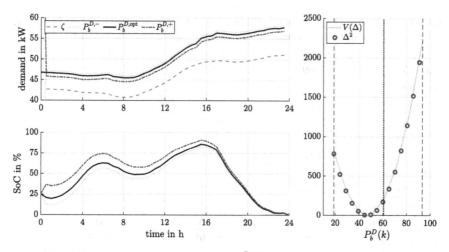

Fig. 7.1 Left: optimal aggregated power demand $P_b^{D,\text{opt}}$ with respect to peak-shaving and associated average SoC x (solid black lines). The dotted blue line and dash-dot red line correspond to minimal and maximal aggregated power demand, $P_b^{D,-}$ and $P_b^{D,+}$, at the first time instant, respectively. For better readability, the power demand at time instant $k = 0$ is only indicated; the values are listed in Table 7.3. Right: exact costs of flexibility versus approximation. The vertical dotted black line marks the current net consumption $W(k)$. The absolute error is listed in Table 7.2

Table 7.2 Absolute error $\left|\Delta^2 - V(\Delta)\right|$ of the quadratic approximation of the costs depicted in Fig. 7.1 (right)

Δ	-28.0	-17.5	-7.1	0	3.4	13.8	24.8	35.7	46.7
Δ^2	782.0	306.9	50.0	0	11.4	191.0	613.5	1276.0	2178.2
$V(\Delta)$	817.1	319.8	51.7	0	7.3	199.2	624.2	1299.5	2219.3
Error	35.1	12.9	1.6	0	4.1	8.2	10.6	23.5	41.1

Figure 7.1 (right) also confirms that $\Delta \mapsto \Delta^2$ suffices as approximation of the costs of flexibility if the batteries are neither full nor empty at the beginning. Here, we compare the actual costs (solid blue line) with penalizing the quadratic deviation of the active power demand from the optimal value at the current time instant (red circles). Moreover, the absolute error of the quadratic approximation is listed in Table 7.2.

The data communicated to the DSO level contains the lower and upper bound on aggregated active power demand $P_b^{D,-}$ and $P_b^{D,+}$, respectively, as well as the optimal demand $P_b^{D,\text{opt}}$ for three MGs. In order to fit the data of the MGs to the IEEE 33-bus system we scale all values such that the power demand without batteries in Scenario (A), $W(k)$, approximately meets the active power demand at the corresponding nodes in the IEEE 33-bus system, see Fig. 7.2 for the location of the MGs. Alternatively, one could adjust the number of residential energy systems (taken from the Australian DSO's data) in order to meet the values in the IEEE 33-bus system. However, this

Table 7.3 Flexibility data provided by MGs consisting of lower and upper bound for aggregated active power demand and the optimal demand associated with the peak-shaving OCP (3.3)

	MG 1	MG 2	MG 3
IEEE 33-bus	60.00	420.00	210.00
Scenario (A)			
$W(k)$	60.60	419.68	210.16
$P_b^{D,+}(k)$ (max.)	104.41	742.00	382.88
$P_b^{D,\mathrm{opt}}(k)$ (opt.)	47.81	331.99	170.22
$P_b^{D,-}(k)$ (min.)	18.82	113.06	56.22
Scenario (B)			
$W(k+1)$	50.20	464.83	187.91
$P_b^{D,+}(k+1)$ (max.)	94.01	787.15	360.63
$P_b^{D,\mathrm{opt}}(k+1)$ (opt.)	47.68	335.00	170.08
$P_b^{D,-}(k+1)$ (min.)	8.29	162.95	33.97

For simplicity scenarios (A) and (B) are assumed to occur within consecutive time steps. All values are in kW, and have been scaled to be in line with the values at nodes 16, 24 and 32 of the IEEE 33 bus system

additional effort would qualitatively yield the same outcome. Therefore, we choose to simply scale the values accordingly which can be interpreted as increasing/decreasing demands as well as storages within each MG. The values are listed in Table 7.3. Note that the P values for scenario (B) depend on the choices made in scenario (A), in particular on the resulting SoCs. These choices are (slightly) affected by the methods used on the DSO and TSO level in both (A). The values for (B) in Table 7.3 are based on using the surrogate model in combination with SDP in (A); the values resulting from other combinations are given in the appendix.

Based on this data, the DSO knows the range of active power demand that can be achieved by manipulating the local power demands via battery control. Furthermore, costs can be associated with the flexibility by computing the deviation from the optimal demand.

7.2 Flexibility of the Distribution Grid

We use a modification of the IEEE 33-bus distribution grid [2] to represent the distribution grid. It is shown in the left graph in Fig. 7.2. Bus 1 is the reference generator which represents the transmission grid, and buses 16, 24, and 32 represent the three microgrids. As mentioned, we only cluster the non-microgrid loads for the sake of simplicity. Figure 7.2 also shows the identified clusters of load buses (middle) and the resulting surrogate model obtained by clustering the load nodes in these clusters (right). The surrogate model contains a loop because the fifth cluster

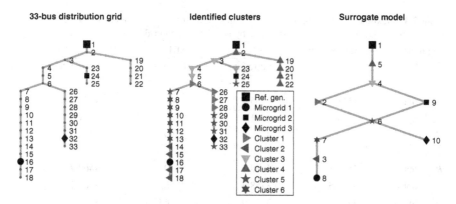

Fig. 7.2 The 33-bus distribution grid used as test instance (left), the identified load node clusters (middle), and the surrogate model (right) obtained with the identified clustering

(bus 6 in the surrogate model) contains buses from several of the branches in the original grid. In fact, it is desirable to retain the tree structure of the original grid because specialized analysis methods exist for such grids. However, this requires *constrained* clustering methods which are outside the scope of this work.

Given the minimal, maximal, and optimal active microgrid power demands (see Table 7.3 as well as Table A.1 in the appendix), we compute the corresponding minimal, maximal, and optimal active power demand of the entire distribution grid as described in Chap. 4 for both scenarios (A) and (B). We compare the results obtained with the original full-order model and the surrogate model (both grids are shown in Fig. 7.2), and we compare the solutions obtained using SDP with those obtained using MATPOWER.

The results are shown in Table 7.4. In general, the results for MATPOWER and SDP are very similar, with slightly lower objective values in the case of SDP, which can be explained by the nature of this relaxation. However, significant numerical problems arise in the computation of the upper bound (maximization) in the SDP case. These issues make the results for this subproblem uncomparable to the MATPOWER results. There are three underlying factors that contribute to the problem. First, the SDP relaxation omits the non-convex rank constraint, which removes some dependencies in the voltage variables. Second, the 33-node IEEE instance does not contain thermal limits for lines. Third, the maximization objective results in a solution that is supposed to spend as much energy as possible. As a result, the 'optimal' solution is to waste amounts of energy that are significantly higher than the demand. The maximum possible output of the generator is spent to overheat the lines, which is connected to the thermal limit being unavailable. Note, that even if the thermal limit is available in a different instance, the maximization objective still results in unreasonable amounts of energy being 'dumped' to overheat the lines. This, in fact, can also be true for nonlinear solvers such as MATPOWER. Our solution in this hierarchy

Table 7.4 Maximum $(P_{b_{ref}}^u)$, optimal $(P_{b_{ref}}^{opt})$, and minimum $(P_{b_{ref}}^\ell)$ amount of power (in kW) required from node 5 of the transmission grid by the distribution grid

Approach	MATPOWER		SDP	
Model	Original	Surrogate	Original	Surrogate
Scenario (A)				
$P_{b_{ref}}^u$ (max.)	4518.7	4455.4	4506.9	4448.2
$P_{b_{ref}}^{opt}$ (opt.)	3764.1	3761.9	3764.1	3761.8
$P_{b_{ref}}^\ell$ (min.)	3369.3	3411.0	3369.3	3411.0
Scenario (B)				
$P_{b_{ref}}^u$ (max.)	4527.8	4468.1	4519.2	4460.7
$P_{b_{ref}}^{opt}$ (opt.)	3767.5	3758.3	3768.0	3764.7
$P_{b_{ref}}^\ell$ (min.)	3385.5	3427.9	3386.9	3428.3

is to introduce additional constraint that limits the power loss to 8% of power generation in the system. This value was estimated from the two minimization subproblems (for optimal and minimal energy consumption) for this particular instance.

Using the surrogate model, the range of admissible active powers, $[P_{b_{ref}}^\ell, P_{b_{ref}}^u]$, is contained within that obtained with MATPOWER as intended. Furthermore, the optimal value $(P_{b_{ref}}^{opt})$ is within ± 10 kW. Finally, the differences in the results obtained in scenarios (A) and (B) are fairly similar. The reason is that some batteries in the microgrids are discharged and some are charged (see Fig. 7.4 in Sect. 7.5). Additionally, the optimal active powers do not change significantly with the state of charge (see Fig. 7.1).

In conclusion, the lower bound and the optimal value obtained using SDP are close to those obtained using MATPOWER. Furthermore, using the surrogate model together with MATPOWER leads to conservative and physically realizable active powers which can be used in the optimal power flow computations for the transmission grid described in the following section.

7.3 Flexibility in the Transmission Grid

We consider the IEEE 9-bus system as the TSO grid, see Fig. 7.3, with node 5 as the flexible demand load (the connection to the single 33-node DSO grid). We first find the original OPF solution, i.e. without considering flexibility at node 5, and without security or stability. This solution is reported in the column labeled "original" of Table 7.5. Then, we solve the security-constrained OPF problem of Sect. 5.1 with flexibility at node 5. We solve this problem both with MATPOWER's Interior Point Solver (MIPS) [13, 14], and the SDP formulation. The results are shown under the column labeled "flexible" of Table 7.5. Introducing additional constraints corresponding to security increases the optimal value due to higher power demand

Table 7.5 Value of objective function (5.1) ($/h) and demand at flexible node (real power, MW) for the optimal solution of the TSO grid

Approach	MATPOWER			SDP		
Model	Original	Flexible	Stability	Flexible	Security	Stability
Scenario (A)						
Objective	5301.11	5290.9	8045.3	5290.9	5472.0	8044.2
Flexible demand	90	89.54	80.56	89.54	89.53	80.56
Scenario (B)						
Objective	5301.11	5290.9	7894.6	5290.9	5472.0	7882.9
Flexible demand	90	89.54	80.88	89.54	89.53	80.90

necessary to fulfill the constraints. Moreover, we can see in Table 7.5 that scenario (A) results in a slightly higher flexible demand than scenario (B) and that the solutions for the flexible network were exactly the same for MATPOWER and our SDP approach. As expected, the relaxation provides objective function values that are slightly lower than the fully constrained model of MATPOWER. In conclusion, using SDP relaxation in this instance results in solutions that are reasonably close to the full nonlinear problem, even for a network that does not have a tree structure. Imposing additional restrictions (security and stability) gives us higher cost, which is a trade-off for including preventive measures in the model.

Next, we solve the stability problem of Sect. 5.2 with node 5 considered a flexible load, both with MATPOWER's MIPS solver and the SDP formulation. The right-hand grid of Fig. 7.3 is the obtained dynamic structure-preserving model as detailed in Sect. 2.3. The p.u. transient reactances of the machines are specified as $x_1 = 0.0608$, $x_2 = 0.1198$ and $x_3 = 0.1813$, taken from Example 2.6 from [1]. We take the desired worst-case dissimilarity to be $\bar{\gamma} = \frac{\pi}{27} \approx 0.1164$. Algorithm 1 only loops once, producing the sequence $\gamma_0 = 0.1412$, $\gamma_1 = 0.1074$. See the column labeled "stability" of Table 7.5 for the results. As can be seen, forcing the power-flow set-point to result in coercive generators results in a higher objective function value, and decreases the active power demanded from the flexible load. Since the results at the DSO level do not differ significantly between scenarios A and B, the results at the TSO level are also almost exactly the same.

At this point we choose the power-flow set-point on the TSO grid determined from the stability-guaranteed power-flow problem, and fix the power demand at the flexible load, node 5.

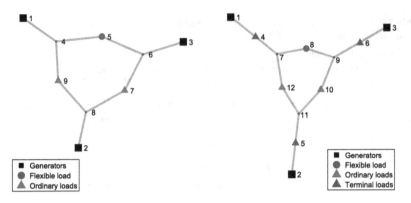

Fig. 7.3 IEEE 9-bus test instance network (left) and the corresponding structure preserving (SP) model (right)

7.4   Implementation at the Distribution Grid Level

Once the optimal power-flow computations of the transmission grid have been completed, the total active power delivered to the distribution grid is fixed. Here, we solve the optimization problem described in Sect. 6.1 in order to determine the optimal active powers supplied to each of the microgrids. As mentioned in Sect. 7.2, the minimal, maximal, and optimal active power demands of the microgrids are scaled in the computations related to the distribution grid.

Table 7.6 shows the active power allocated to each microgrid. As in Sect. 7.2, we show the values obtained with both the original full-order model and the surrogate model. Furthermore, we compare the results obtained using SDP and MATPOWER.

In scenario (A), the surrogate model slightly underestimates the line power losses, because the difference in the amount of power sent to the microgrids is more than the difference in power delivered by the transmission grid. The reverse is true in scenario (B), where the losses are overestimated as more additional power is delivered from the transmission grid, while microgrid loading changes less.

The results obtained using SDP are very similar to the ones from MATPOWER, but the amount of power delivered to the microgrids is slightly lower in scenario (A), which results in a significantly lower objective function value (as it is described as the square of difference from the optimum). On the other hand, in scenario (B), the values for both the microgrid loading and the objective function are much closer due to less power being available in the system. The difference in the magnitude of the objective function value between scenario (A) and scenario (B) can also be attributed to the structure of the objective function, as the loading of the microgrids is far away from the optimum in scenario (B).

Table 7.6 The amount of active power supplied by the transmission grid (kW), the optimal amounts of active power delivered to the microgrids from the distribution grid (kW), and the corresponding value of the objective function (6.1a) ($/h)

Approach	MATPOWER		SDP	
Model	Original	Surrogate	Original	Surrogate
Scenario (A)				
Transmission grid	3874.7	3922.6	3874.7	3922.6
Microgrid 1	76.9	96.2	62.8	88.6
Microgrid 2	358.5	379.2	348.5	373.7
Microgrid 3	198.8	218.0	186.1	211.9
Objective	2366.3	6853.3	748.3	5138.5
Scenario (B)				
Transmission grid	3385.5	3427.9	3386.9	3428.3
Microgrid 1	8.3	8.3	8.3	8.3
Microgrid 2	164.6	192.3	166.0	192.2
Microgrid 3	34.0	34.0	34.0	34.4
Objective	49,239.0	40,396.8	48,859.4	40,341.5

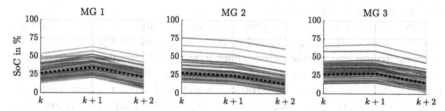

Fig. 7.4 Implementation at MGs. At time k the batteries start with initial SoCs x_i^k. First (scenario (A)), most of the batteries are charged in order to store the surplus of active power from the upper levels. Then (scenario (B)), the energy is set free in order to reduce the overall demand. The dotted black line depicts the average SoC within each MG

7.5 Implementation at MGs

The implementation at the batteries for both Scenarios (A) and (B) is visualized in Fig. 7.4. First, we store the surplus of energy by charging the batteries on average. Then, the energy is set free in order to reduce the overall active power demand.

Note that in our setting, i.e., starting with $x_i^k \sim \mathcal{N}(0.5, 0.05)$ and only considering two consecutive time steps, the batteries are neither fully charged nor fully discharged. In the long run, however, one might be interested in always having a certain amount of flexibility available in case of emergency. To this end, the bounds in (3.2a) might be tightened in order to keep the average SoC within a reasonable range. Note that this only affects the optimal solution of (3.3). The total amount of flexibility that is communicated to the DSO stays the same. Therefore, the buffer can be exploited by the upper levels. Alternatively, one might introduce *tube constraints* [4, 9] based on predictions to ensure that the batteries can be charged whenever a surplus of energy is expected and discharged otherwise.

7.6   Conclusion and Outlook

Our numerical simulations show that the presented approach works as intended: First, each MG computes the amount of inherent flexibility, associates costs, and propagates this information to the DSO. Then, the DSO utilizes this data to solve OPF problems using both semi-definite programming and cluster-based surrogate models and communicates the results to the TSO. On the highest level, the optimal power for security-constrained OPF on the one hand or guaranteeing synchronized generators on the other hand is determined. Then, the decision is communicated downwards through the entire grid hierarchy to the MGs, where the resulting charging signal is implemented at the residential batteries. In particular, if the current demand falls below the desired value, e.g. due to high generation via renewables, the superfluous energy is stored in the residential batteries; otherwise previously stored energy is set free to reduce the overall demand.

Recently, purely data-driven approaches based on Willems et al.'s so-called fundamental lemma [12] were proposed to construct optimal and predictive control schemes. To this end, persistently-exciting input-output trajectories were used to derive a non-parametric description of the system behavior, see, e.g., [3, 5] as well as [6] and the references therein. Based on an extension to descriptor systems [10] these techniques were tailored to power networks [11] such that the previously exploited modelling step can be skipped to optimize the system behaviour in data-based manner. Such data-driven techniques might be implemented within each level of the grid hierarchy and, thus, be used to improve the performance and reduce the computational burden of the presented approach.

References

1. Anderson PM, Fouad AA (2003) Power system control and stability, 2nd edn. IEEE Press
2. Baran ME, Wu FF (1989) Network reconfiguration in distribution systems for loss reduction and load balancing. IEEE Trans Power Delivery 4(2):1401–1407
3. Berberich J, Köhler J, Müller MA, Allgöwer F (2021) Data-driven model predictive control with stability and robustness guarantees. IEEE Trans Autom Control 66(4):1702–1717
4. Braun P, Faulwasser T, Grüne L, Kellett CM, Weller SR, Worthmann K (2018) Hierarchical distributed ADMM for predictive control with applications in power networks. IFAC J Syst Control 3:10–22
5. De Persis C, Tesi P (2020) IEEE Trans Autom Control 65(3):909–924
6. Markovsky I, Dörfler F (2021) Behavioral systems theory in data-driven analysis, signal processing, and control. Ann Rev Control 52:42–64

7. Meinecke S, Sarajlić D, Drauz SR, Klettke A, Lauven L-P, Rehtanz C, Moser A, Braun M (2020) SimBench—a benchmark dataset of electric power systems to compare innovative solutions based on power-flow analysis. Energies 13(12):3290

8. Ratnam EL, Weller SR, Kellett CM, Murray AT (2015) Residential load and rooftop PV generation: an Australian distribution network dataset. Int J Sustain Energy

9. Sauerteig P, Worthmann K (2020) Towards multiobjective optimization and control of smart grids. Optim Control Appl Methods (OCAM) 41:128–145

10. Schmitz P, Faulwasser T, Worthmann K (2022) Willems' fundamental lemma for linear descriptor systems and its use for data-driven output-feedback MPC. IEEE Control Syst Lett 6:2443–2448

11. Schmitz P, Engelmann A, Faulwasser T, Worthmann K (2022) Data-driven MPC of descriptor systems: a case study for power networks. IFAC-PapersOnLine. Preprint: arXiv:2203.02271

12. Willems JC, Rapisarda P, Markovsky I, De Moor BLM (2005) A note on persistency of excitation. Syst Control Lett 54(4):325–329

13. Zimmerman RD, Murillo-Sánchez CE (2020) MATPOWER (version 7.1)

14. Zimmerman RD, Murillo-Sánchez CE, Thomas RJ (2011) MATPOWER: steady-state operations, planning, and analysis tools for power systems research and education. IEEE Trans Power Syst 26(1):12–19

Appendix
Complete Flexibility Results for the Microgrids

Here, we present the results for scenario (B) where both MATPOWER and SDP are used with the original and the surrogate model for the computations in scenario (A). They are shown in Table A.1. As is evident, the results for scenario (B) are unaffected by whether the original model or surrogate model is used. The reason is that the optimal active power demand as well as the corresponding upper and lower bounds only change very little depending on the state of charge. This can also be seen in Fig. 7.4 (left) where the state of charge varies from 25% to around 60% during the first 6 h, but the optimal power demand is almost unchanged.

Table A.1 Lower bound, upper bound, and optimal value of the aggregate active power demand of the microgrids in scenario (B) based on the results for scenario (A)

Approach	MATPOWER		SDP	
Model	Original	Surrogate	Original	Surrogate
Microgrid 1				
$W(k+1)$	50.20	50.20	50.20	50.20
$P_b^{D,+}(k+1)$ (max.)	94.01	94.01	94.01	94.01
$P_b^{D,\mathrm{opt}}(k+1)$ (opt.)	47.80	47.60	47.94	47.68
$P_b^{D,-}(k+1)$ (min.)	8.29	8.29	8.29	8.29
Microgrid 2				
$W(k+1)$	464.83	464.83	464.83	464.83
$P_b^{D,+}(k+1)$ (max.)	787.15	787.15	787.15	787.15
$P_b^{D,\mathrm{opt}}(k+1)$ (opt.)	335.18	334.93	335.31	335.00
$P_b^{D,-}(k+1)$ (min.)	164.58	162.49	166.11	162.95
Microgrid 3				
$W(k+1)$	187.91	187.91	187.91	187.91
$P_b^{D,+}(k+1)$ (max.)	360.63	360.63	360.63	360.63
$P_b^{D,\mathrm{opt}}(k+1)$ (opt.)	170.25	170.00	170.43	170.08
$P_b^{D,-}(k+1)$ (min.)	33.97	33.97	33.97	33.97

All values are in kW

© The Author(s), under exclusive license to Springer Nature Switzerland AG 2023
T. Aschenbruck et al., *Hierarchical Power Systems: Optimal Operation Using Grid Flexibilities*, SpringerBriefs in Energy,
https://doi.org/10.1007/978-3-031-25699-8

Printed in the United States
by Baker & Taylor Publisher Services